The Geneva Act (1999) of the Hague Agreement Concerning the International Registration of Industrial Designs

Drafting History and Analysis

The Geneva Act (1999) of the Hague Agreement Concerning the International Registration of Industrial Designs

Drafting History and Analysis

by
William T. Fryer, III

KLUWER LAW
INTERNATIONAL

Published by:
Kluwer Law International
P.O. Box 85889
2508 CN The Hague
The Netherlands

Sold and distributed in North, Central and South America by:
Aspen Publishers, Inc.
7201 McKinney Circle
Frederick, MD 21704
USA

Sold and distributed in all other countries by:
Extenza-Turpin Distribution Services
Stratton Business Park
Pegasus Drive
Biggleswade
Bedfordshire SG18 8TQ
United Kingdom

A C.I.P. catalogue record for this book is available from the Library of Congress.

Printed on acid-free paper

Typeset by *Steve Lambley Information Design*, The Hague.

ISBN 90-411-2117-X

Acknowledgments

The experience of writing a book is humbling. So many people contributed to the process, and provided enlightenment along the way. It would be impossible to mention them all. I thank them for their support and encouragement.

This book describes the leadership of the World Intellectual Property Organization (WIPO) and the contributions of many people who participated in development of The Geneva Act (1999) of the Hague Agreement on the International Registration of Industrial Designs. I am greatly appreciative to WIPO for allowing me to use in this book WIPO documents related to the 1999 Act.

Over the more than ten years of my participation in the 1999 Act, development and preparation of this book, my wife, Mary, and daughters have been very encouraging and patient as I attended international meetings, conducted research and committed to preparing this book. The University of Baltimore School of Law, Baltimore, Maryland, U.S., supported this project in many ways.

Foreword

The development of the Geneva Act (1999) of the Hague Agreement Concerning the International Registration of Industrial Designs took place in seven experts meetings from 1991 to 1997 and the diplomatic conference was in 1999. The treaty became operational on April 1, 2004. The author of this book attended each of these meetings as the representative of the American Bar Association, Section of Intellectual Property Law (ABA). On July 7, 2004 the author of this book attended a briefing session for users of The Geneva Act (1999). During this treaty development the author of this book published several articles analyzing progress made at the meetings of experts.

This book analyzes the drafting history of the Geneva Act (1999) (also identified as the 1999 Act), and introduces the treaty operation, with insight gained from having participated in these meetings. In this book's appendices are several important resource documents that provide historical depth and practical insight into the Geneva Act (1999). These materials include the Geneva Act (1999) text, common regulations, administrative instructions, and earlier Hague Agreement Acts of 1934, 1960, 1961 and 1967. A bibliography is provided of the principal documents used at the meetings of experts and the diplomatic conference, as well as articles published on the Hague Agreement.

There are several ways to use this book, depending on the reader's background and interest. If a reader wants a good foundation on the 1999 Act, the place to start reading is the General History chapter. An understanding of the historical framework for the treaty is very important. The next part to read is the Operation – Introduction chapter, with its drawings that depict the relationships and timing for the major types of national design registration systems and the 1999 Act. There will be questions left after reading this introductory material, and the reader is referred to the core of this book, the Topics chapter. In these topics, alphabetically arranged, there is an in-depth review of the basic 1999 Act features. On many subjects there are several interrelated topics that must be understood and the reader should consult all relevant topics for a comprehensive understanding of the 1999 Act.

Another way to use this book is to review the subject of interest in the Topics chapter and refer to the Operation – Introduction chapter if there are questions on relationships or timing.

For government officials studying the 1999 Act for preparation of documents to adhere to the 1999 Act, a very useful digest has been included in the Topics chapter, under the heading "Declarations". This digest lists all the declarations and notifications that a country or intergovernmental organization may need to make to WIPO and the International Bureau to commence operation under the treaty. The task of interfacing a national design registration system with the 1999 Act is a very important step. While the policy of the 1999 Act development was to minimize the need to change national design registration laws to adhere to the treaty, there are several requirements and many optional provisions that may be essential to incorporate.

As the 1999 Act adds new members, there will be additional information needed to prepare International Registration applications. Each new contracting party will submit declarations and notifications that will control the procedures and legal effect with respect to the treaty. This information will be on the WIPO website for the Hague Agreement System at the URL http://www.wipo.int. An updated International Registration application form will be available on the WIPO website when new contracting parties are added. Therefore, it is essential that users regularly check the WIPO web site, and particularly before filing an International Registration application.

The author welcomes suggestions and comments to make this book more useful. These communications can be sent to the author's e-mail address at fryer@fryer.com.

Contents

Bibliography

WIPO (Geneva) Publications

The Hague Agreement concerning the International Deposit of Industrial Designs of November 6, 1925: Regulations (As in force on January 1, 1998) and Administrative Instructions (as in force on January 1, 1998), Publication, No. 262(E) (1983).

Meeting of Experts on the Development of the Hague Agreement concerning the International Deposit of Industrial Designs, records at WIPO (1991-1998).

The First Twenty Five Years of the World Intellectual Property Organization, from 1967 to 1992, Profile: The Hague Agreement, International Bureau of Intellectual Property, Geneva (1992).

The Hague Agreement concerning the International Deposit of Industrial Designs: Objectives, Main Features, Advantages, WIPO Publication, NO 419(E) (1998).

Industrial Designs and the Hague Agreement: An Introduction, WIPO Publication, NO 429(E) (1999).

The Geneva Act (1999) of the Hague Agreement concerning the International Registration of Industrial Designs: Main Innovations Compared with the Present System Provided for in the 1934 and 1960 Acts of the Agreement, WIPO Publication No. 453(E) (1999).

Diplomatic Conference on the Geneva Act (1999) of the Hague Agreement concerning the International Registration of Industrial Designs, Records at WIPO, including transcript of discussion and action on final treaty text (1999).

Records of the Diplomatic Conference for the Adoption of a New Act of the Hague Agreement Concerning the International Deposit of Industrial Designs (Geneva Act) June 16 to July 6, 1999 WIPO (2002).

International Classification for Industrial Designs (Locarno Classification), 8th edition, Publication No. 501(E) (2003).

Hague Agreement Concerning the International Registration of Industrial Designs, Acts of 1925, 1934, 1960, 1961, 1967 (as amended September 28, 1979), Geneva Act 1999, Common Regulations (as in force on April 1, 2004), and Administrative Instructions (as in force on April 1, 2004), WIPO Publication 269(E) (2004).

Guide to the International Registration of Industrial Designs under the Hague Agreement, Publication No. 857(E) (2004)

WIPO Documents – Geneva Act Preparation, Meeting of Experts and Diplomatic Conference

Special Notes

1. This list contains selected, principle documents from the Official list published by WIPO. Unless otherwise stated, all the documents listed were in English, and that identification in the documentation identification has been omitted.
2. The introductory title for each meeting of experts document had a common part, unless otherwise indicated. It was: "Meeting of Committee of Experts on the Development of the Hague Agreement Concerning the International Deposit of Industrials." On the list the document citation was shorted to identify the meeting and document number.
3. Only the Seventh Meetings of Experts documents and all the Diplomatic Conference documents were accessible from the WIPO web site (http://www.int) News & Information Resources Meetings page under the Meeting heading (Archived meeting documents), as of July 5, 2004.

First Session, Meeting of Experts (April 8-11, 1991)

1. Agenda – WIPO H/CE/I/1 (Feb. 15, 1991)
2. Possibilities of Improving the System of International Deposit of Industrial Designs – WIPO H/CE/I/2 3. (Feb. 15, 1991)
4. WIPO H/CE/I/INF/1 (date April 8, 1991), Information on Certain Provisions in National Laws for the Protection of Industrial Designs (Chart showing features of industrial Designs system for 19 countries).
5. Report adopted by the Committee of Experts – WIPO H/CE/I/3 (April 10, 1991)

Second Session, Meeting of Experts (April 27-30, 1992)

1. Agenda – WIPO H/CE/II/1
2. Draft Treaty on the International Registration of Industrial Designs – H/CE/II/2
3. Report adopted by the Committee of Experts – H/CE/II/3

Third Session, Meeting of Experts (April 26-30, 1993)

1. Agenda – WIPO H/CE/III/1
2. Draft New Act – H/CE/III/2 (Feb. 26, 1993)
3. Report adopted by the Committee of Experts – H/CE/III/3 (April 30, 1993)

Fourth Session, Meeting of Experts (Jan. 31to Feb. 4, 1994)

1. Agenda – H/CE/IV/1 (Nov. 24, 1993)
2. Draft New Act of the Hague Agreement – H/CE/IV/2 (Nov. 24, 1993)
3. Draft Article 9/2 (Proposed by the Delegation of Italy) (Jan 31, 1994)
4. Report adopted by the Committee of Experts – H/CE/IV/4 (Feb. 4, 1994)

Fifth Session, Meeting of Experts (June 13-16, 1995)

1. Agenda – H/CE/V/1 (Feb. 28, 1995)
2. Draft New Act of the Hague Agreement (Feb. 28, 1995)
3. Art. Z – Relations with the TRIPS Agreement, proposed by the Delegation of Switzerland
4. Art. 9(3) – Informal proposal by the European Commission, June 9, 1995
5. Report adopted by the Committee of Experts – H/CE/V/4 (June 16, 1995)

Sixth Session, Meeting of Experts (Nov. 4-8, 1996)

1. Agenda – H/CE/VI/1 (Sept. 6, 1996)
2. Draft New Act of the Hague Agreement – H/CE/VI/2 (Sept. 6, 1996)
3. Notes on the Draft New Act of the Hague Agreement – H/CE/VI/3 (Sept. 6, 1996)
4. Draft Rules under the Draft New Act of the Hague Agreement – H/CE/VI/5 (Nov. 7, 1996)
5. Report adopted by the Committee of Experts – H/CE/VI/5 (Nov. 7, 1996)

Seventh Session, Meeting of Experts (Nov. 3-7, 1997)

1. Agenda – H/CE/VII/1 (Sept. 3, 1997)
2. Draft New Act of the Hague Agreement, H/CE/VII/1 (Sept. 3, 1997)
3. Notes on the Draft New Act of the Hague Agreement, H/CE/VII/3 (Sept. 3, 1997)
4. Draft Regulations under the Draft New Act of the Hague Agreement, H/CE/VII/4 (Sept. 3, 1997)
5. Notes on the Draft Regulations under the Draft New Act of the Hague Agreement, H/CE/VII/5 (Sept. 17, 1997)
6. Report adopted by the Committee of Experts – H/CE/VII/6 (Nov. 7, 1997)

Final Draft Act Preparation for the Diplomatic Conference

1. Draft New Act of the Hague Agreement – H/CE/3 Prov (June 30, 1998)
2. Draft Regulations under the Draft New Act of the Hague Agreement – H/DC/4 Prov. (June 30, 1998)

Diplomatic Conference for the Adoption of a New Act of the Hague Agreement Concerning the International Deposit of Industrial Designs (Geneva, June 16 to July 6, 1999)

1. Draft Agenda – H/DC/1 (Dec. 15, 1998)
2. Basic Proposal for the New Act of the Hague Agreement – H/DC/3 (Dec. 15, 1998)
3. Basic Proposal for the Regulations under the New Act of the Hague Agreement – H/DC/4 (Dec. 15, 1998)
4. Notes on the Basic Proposal for the New Act of the Hague Agreement – H/DC/5 (Dec. 15, 1998)

5. Notes on the Basic Proposal for the Regulations under the New Act of the Hague Agreement – H/DC/6 (Dec. 15, 1998)
6. Geneva Act of the Hague Agreement … Regulations Under the Geneva Act and Agreed Statements by the Diplomatic Conference – H/DC/40 (July 2, 1999)

Books

Cohen, Denis, *The International Protection of Designs* (Hague, Kluwer Law International 2000)

Gervais, Daniel, *The TRIPS Agreement: Drafting History and Analysis* (2nd edition, London, Sweet & Maxwell 2003).

Leaffer, Marshall, 'Industrial Design', 485, *International Treaties on Intellectual Property*, 2nd edition (1997).

U.S. Patent and Trademark Office, Manual of Patent Examining Procedure (8th edition, Revision 1 2003).

Articles

Adams, John, 'International News – WIPO Commentary – Industrial Designs – Recent Developments and Draft Treaty', 4 *E.I.P.R.* D-80 (1992).

Curchod, Francois, 'The Revision of the Hague Agreement Concerning the International Deposit of Industrial Designs', 24 *AIPLA Quarterly J.* 599 (1996).

Fryer, III, William T., 'Report on Hague Agreement (Industrial Designs) Second Meeting of Experts, Held April 27-30, 1992, at WIPO in Geneva', 74 *J. Patent & TM Office Society* 923 (1992).

Fryer, III, William T. Fryer, 'More Bang for Your Design Protection Money: A Report on The Hague Agreement Third Meeting of Experts, July 1993', xxx *J. Patent & TM Office Society* xxx (1993).

Fryer, III, William T., 'Seeking a Benefits Balance in the Industrial Design Treaty Revision (Hague Agreement); Fifth Meeting of Experts, Held June 13-16, 1995', 77 *J. Patent & TM Office Society* 931 (1995).

Fryer, III, William T., 'International Industrial Design Protection Improvement: The Hague Agreement Revision', 2 *U. Balt. Intell. Prop. L.J.* 37 (1993).

Knight, G. Lloyd, 'It's Time to Stop Discriminating Against PCT/USA National Phase Patents', *J. Patent & TM Office Society* 385-400 (1997).

Myles Jelf, 'Grand Designs? The New Registered Designs Regime', 118 *Copyright World* (2002).

Phillips, Jeremy, 'International Design Protection: Who Needs It?', *E.I.P.R.* 431 (1993).

Zanetti, Giulio Cesare, 'A New Treaty for the International Protection of Industrial Designs', 98 *Copyright World* 26 (2000).

Sources Cited in Text

Geneva Act (1999) Articles

Geneva Act (1999) Rules

Geneva Act (1999) Administrative Instructions

Court Decisions Cited in Text

U. S. Statutes Cited in Text

U.S. Public Laws

Publications

CHAPTER I

General History of the Geneva Act (1999), Hague Agreement Concerning the International Registration of Industrial Designs

Meetings of Experts: 1991, 1992, 1993, 1994, 1995, 1996, 1997

Diplomatic Conference: 1999

Overview

The purpose of this chapter is to give a historical context for understanding the Geneva Act (1999) development. There were major events, including broad political issues that gave the Hague Agreement revision a major role in the evolution of intellectual property treaties.

The preparation of the Geneva Act (1999) began with a general consensus that there was a need for expanding the potential membership of the existing Hague Agreement, and thereby creating a global treaty for centralized administration of industrial design application filing and related aspects. The exact form the new treaty would take was not known. Many differences existed in the types of industrial design systems around the would, making this effort a challenge, and requiring several meetings to work out the details.

The principle documentation for each of the preparatory Meeting of Experts and the Diplomatic Conference is listed in the Bibliography.

Many topics were discussed during the Meetings of Experts and the Diplomatic Conference that are now key features of the Geneva Act. The Topics chapter explains in detail the history and use of many of these features and provides references to source documents.

FIRST SESSION OF COMMITTEE OF EXPERTS ON THE DEVELOPMENT OF THE HAGUE AGREEMENT CONCERNING THE INTERNATIONAL DEPOSIT OF INDUSTRIAL DESIGNS, HELD AT WIPO, GENEVA, SWITZERLAND, APRIL 8-11, 1991.

Summary

1. WIPO organized the First Session of the Committee of Experts on the Development of the Hague Agreement Concerning the International Deposit of Industrial Designs (Hague Agreement) under the following governing assembly mandate: "The International Bureau will convene a working group, consisting of representatives of States member of the Hague Union and, as observers, of other States, Intergovernmental Organizations and Non-Governmental Organizations, to consider the advisability of revising the Hague Agreement, or adopting to it a protocol, in order to introduce in the Systems further flexibility and other measures encouraging States not yet part to it to adhere to the Hague Agreement and making it easier to use by applicants." [Program and Budget of WIPO for the 1990-1991 biennium (document AB/XX.2, Annex A, page 42, Item REG.03(31)].

2. The First Session was attended by a significant number of representatives [8 members of the Hague Union, 13 observer States, 3 observer international organizations and 11 non-governmental organizations].

3. The discussion centered on current Hague Agreement features and the need to achieve compatibility with national industrial design laws. The meeting gave representatives an opportunity to become familiar with the Hague Agreement history and operation. Many topics were identified as needing further discussion. A second Meeting of Experts was planned for this purpose.

SECOND SESSION OF COMMITTEE OF EXPERTS ON THE DEVELOPMENT OF THE HAGUE AGREEMENT CONCERNING THE INTERNATIONAL DEPOSIT OF INDUSTRIAL DESIGNS, HELD AT WIPO, GENEVA, SWITZERLAND, APRIL 27-30, 1992.

Summary

1. The Second Session had a noticeable increase in attendance [11 members of the Hague Union, 2 international organizations, and 19 non-governmental organizations].

2. WIPO prepared a Draft Treaty for an International Industrial Design Registration to obtain protection in member countries, and this document set the stage for

many important discussions. These discussions started the process for developing solutions by consensus, a procedure used throughout the experts meetings and the Diplomatic Conference. The topics discussed most extensively at this session were:

(a) Direct filing with WIPO by an applicant was the procedure proposed by the Draft Treaty. The question of filing indirectly, through a national office, was raised and proposed for continued discussion at the next session.

(b) It was made clear that the Hague Agreement revision was primarily to improve international filing and not to create substantial industrial design law harmonization. This point had to be repeated at each expert meeting session and at the Diplomatic Conference. One example of law harmonization briefly discussed and dropped from this treaty revision work was the provision of an unregistered industrial design in combination with registration protection. This combination was under consideration at that time and subsequently adopted by the European Union in its Community Design Regulation.

(c) The deferment of industrial design publication was not proposed in the Draft Treaty. Considerable support was given to adding that feature which is common to most national industrial design registration systems.

(d) It was generally agreed that the short period allowed for refusal of a Hague Agreement International Registration effect under present procedures was a major reason that novelty examination countries had not become Hague Agreement members. It was concluded that this issue should be addressed in the next session.

(e) There was no provision in the Draft Treaty for multiple designs in one International Registration application. The Draft Act did not provide for use of specimens, like cloth samples of two-dimensional design as the basis for establishing a filing date, a feature of the 1960 Act of the Hague Agreement. Addition of these features to the Draft Treaty was strongly supported in the discussion.

Resource: William T. Fryer, III, *Report on Hague Agreement (industrial designs) Second Meeting of Experts, held April 27-30, 1992, at WIPO in Geneva*, 74 J. Patent & TM Office Society 923 (1992).

THIRD SESSION OF COMMITTEE OF EXPERTS ON THE DEVELOPMENT OF THE HAGUE AGREEMENT CONCERNING THE INTERNATIONAL DEPOSIT OF INDUSTRIAL DESIGNS, HELD AT WIPO, GENEVA, SWITZERLAND, APRIL 26-30, 1993.

Summary

1. The discussion intensity increased as very difficult issues were addressed for a second time, and some very constructive solutions were proposed.

2. The Draft Treaty added publication deferment, using a solution proposed by the Swiss Government. It allowed each member to decide whether to have deferment. This proposal was carried through to the final text of the Geneva Act. The textile and fashion industries represented at this session were very pleased with this addition, because it would allow low cost protection for two-dimensional designs through simplified filing procedures.

3. The Draft Treaty continued to allow only direct filing with WIPO of an industrial design registration. The support increased for indirect filing through national offices.

4. The International Registration application contents required to obtain a filing date were major discussion topics. It was clear that more time was needed to work out these details.

5. Another topic discussed was the application requirements for identification of the product that uses the protected design. It was apparent that this provision could impact on the protection scope and may need clarification at the next session.

6. The Draft Treaty proposed a 30-month time period for national offices to refuse an International Registration. Most examination countries considered this time period acceptable, if the details were clarified. The ABA representative questioned whether all rejection grounds would have to be stated in this response. This topic was discussed at later sessions and in the diplomatic conference.

7. The effect of an International Registration was discussed, including prior art effect. It was a topic left to the Diplomatic Conference to resolve.

8. Fee structure was discussed, with considerable emphasis on reduced costs by use of multiple designs in one International Registration application. The question of how to handle these applications in a novelty examination country with a single invention per registration requirement was left for discussion at the next session.

Resource: William T. Fryer, III, *More bang for your design protection money: a report on The Hague Agreement Third Meeting of Experts: July 1993* 76 J. Patent & TM Office Society 91 (1994).

FOURTH SESSION OF COMMITTEE OF EXPERTS ON THE DEVELOPMENT OF THE HAGUE AGREEMENT CONCERNING THE INTERNATIONAL DEPOSIT OF INDUSTRIAL DESIGNS, HELD AT WIPO, GENEVA, SWITZERLAND, JANUARY 31 – FEBRUARY 4, 1994

Summary

1. The fourth Meeting of Experts successfully addressed several important topics and helped identify other issues needing resolution. This session was devoted substantially to meeting the needs of novelty examination countries. Some issues that could only be resolved at the Diplomatic Conference were discussed for the first time at this session. Other important topics were discussed that had to be placed on the next session agenda. The Draft Act title was edited to use the more current term “Registration” instead of the former “Deposit.”

2. The procedures for establishing when a filing date was established were substantially completed. These provisions organized the determination according to a minimum list of application contents to obtain a filing date. A second list of content items was established to allow novelty examination members to continue their special requirements to obtain a filing date. A third list of options allowed content which would not prevent receipt of a filing date. While these provisions seemed relatively easy to apply, they required a lot of careful attention to detail for the necessary result to occur.

3. One of the requests in the Third Session by several parties was provision for indirect filing of an International Registration through national offices. The Fourth Session Draft Treaty provided for indirect filing. Another general request from the Third Session was to extend the minimum period of an International Registration from 10 to 15 years from the International Registration date. This change, substantive in nature with procedural impact also, was a significant accomplishment in expanding industrial design protection.

4. Several new issues were addressed that needed further discussion at the next meeting. The question of unifying national industrial design registration requirements through standards set in the Draft Act was addressed in some detail. A small step forward toward this goal was made, by agreeing to a provision that a national office could not require more than six different views of the industrial design to obtain a filing date.

5. The issue of prior art effect for an International Registration was discussed only briefly and planned for the next session. There were several divergent views on this topic. In the same situation was the topic of whether the European Union would

have a separate vote if it was a member of the treaty. The topic of WIPO recording a change in an International Registration ownership needed more research, and its discussion was delayed until the next session.

6. The Draft Act administrative provisions, included for the first time, were discussed briefly, due to the limited time available. The question of whether intergovernmental organization membership should allow that organization to vote, in addition to the votes of its States, was discussed, and there were several different views. This topic was carried forward to the next session.

FIFTH SESSION OF COMMITTEE OF EXPERTS ON THE DEVELOPMENT OF THE HAGUE AGREEMENT CONCERNING THE INTERNATIONAL DEPOSIT OF INDUSTRIAL DESIGNS, HELD AT WIPO, GENEVA, SWITZERLAND, JUNE 13-16, 1995

Summary

1. The fifth Meeting of Experts was a very difficult session, with most of the concerns over creating an acceptable system being discussed in detail. It was the unavoidable time when countries with simple registration that had no substantive examination before registration found a lack of balance in the Draft Act. In this group's view, there were too many concessions made already to accommodate the examining system countries, particularly Japan and the U.S.. The simple registration system countries wanted more concessions from the novelty examining system countries, to achieve this balance.

2. The non-examining countries' main concerns centered around the delay in when design protection began. The original Hague Agreement system was planned for simple registration systems. They were efficient and low cost. An International Registration application was filed, processed by the International Bureau in a few months, and published to have legal effect in some countries for infringement rights and prior art effect from the International Registration filing date.

3. The Draft Act version considered at the Fifth Session had significant benefits for the textile industry, an emphasis expressed in GATT-TRIPS, art. 25(2). There was the option to have a deferment of publication with a reduced initial filing fee and the full fee payment coming later. Overall, there was satisfaction with the Draft Treaty that industries with many designs could manage to protect their important designs at reasonable costs. The countries that had no experience with deferred publication learned that this procedure was deeply embedded in the design protection process

of many countries. There had to be concession to have the deferment feature in a global design protection system.

4. For the examining countries, the Draft Act attempted to allow use of their current substantive examination systems. The U.S. delegation was not satisfied on the timing for forwarding an indirectly filed International Registration application, due to its unique national security review procedure. The review could delay transmittal of an application to the International Bureau. The Draft Act provision for indirect filing in national offices greatly simplified foreign filing for the U.S., as it incorporated the security review. A short additional time was allowed to compensate for the U.S. national security review. The debate shifted to keeping the delay introduced by the national security review to a minimum.

5. The U.S. proposed a new provision that would allow the U.S. to retain the *In re Hilmer* law on prior art effect, requiring that an application had to be filed in the U.S. to have prior art effect from that date. This U.S. proposal was strongly opposed by most countries. The proposal to allow refusal of an International Registration effect by a country for up to 30 months, supported by most examining national offices was opposed by many countries, creating additional dissatisfaction among the non-examining countries.

6. The concern for prompt design protection was at the center of a forceful discussion over what some delegations considered special treatment for examining countries. One proposal discussed was to allow accelerated examination in national examining officers for International Registrations, ahead of national applications filed earlier. The European Union (EU) delegation pressed this proposal. The U.S. rejected it, but the concern for prompt processing and procedures to accelerated prosecution for all applications was understood. The U.S. Patent and Trademark Office (PTO) expressed a willingness to increase the speed of design substantive examination.

7. On the surface, the meeting participants appeared widely divided on many issues. In fact, these differences were narrowed by the compromises made. The time taken at the meetings of experts to explain the history and operation of different national protections systems was producing good results. Slowly and with considerable reluctance, it became apparent that a global design protection system would have to include some major compromises to accommodate the two general types of design protection systems.. The only question was how to achieve a balanced agreement.

8. WIPO announced a proposal to force the Draft Act work to a conclusion. It proposed to set up in the next Draft Act two chapters, a Chapter I for the simple registration systems with features that all countries could agree on, and a Chapter II that would be used only by examining countries. There would be a provision for

each country that complied with the provisions of Chapter I to deny the right of an examining country to file under the Draft Act, if the examining country required use of a provision in Chapter II. If an examining country was denied the right to file in a country for this reason, the examining country could reciprocate and deny the right to file against that country. The time had arrived for the representatives participating in these meetings to decide if a global system could be set up with examining countries. The next Meeting of Experts would be the deciding event for this project.

Resource: William T. Fryer, III, *Seeking a benefits balance in the Industrial Design Treaty Revision (Hague Agreement); Fifth Meeting of Experts, held June 13-16, 1995*, 77 J. Patent & TM Office Society 931 (1995).

SIXTH SESSION OF COMMITTEE OF EXPERTS ON THE DEVELOPMENT OF THE HAGUE AGREEMENT CONCERNING THE INTERNATIONAL DEPOSIT OF INDUSTRIAL DESIGNS, HELD AT WIPO, GENEVA, SWITZERLAND, NOV. 4-8, 1996

Summary

1. The sixth Meeting of Experts was successful in overcoming many of the frustrations experienced in the Fifth Session, and the spirit at the end was much more positive.

2. The last Meeting of Experts ended with the controversial conclusion that some countries should have the right to deny an examining country's design owners the right to file under the Draft Act, if certain conditions occurred. This provision was in the Sixth Session Draft Treaty. The consequences of this refusal would be to allow reciprocal denial by the country refused the right to file. It was a step backward to the days where conduct between nations was based on reciprocity, rather than agreement on certain principles. The potential was to have two groups of members to a treaty with the threat of refusal held over each other for various reasons.

3. The sixth Meeting of Experts Draft Treaty was organized to accomplish this reciprocity, by establishing Chapters I and II. Chapter I had the basic requirements that would create a simple registration system compatible procedures, including some of the newly agreed to provisions. Chapter II was just for examining members. If an examining member took advantage of some special features in Chapter II, to adapt the Draft Act more effectively to its system, a Chapter I member could deny the right to file in their country by persons in the examining country. For example, if there was a need to extend the time limit when an International Registration could

be refused, that action could subject that examining country to possible exclusion from filing in one or more member countries.

4. The good news was that this Chapter I and II approach was rejected by the delegates. The underlying reasons for the division remained, and the resultant negotiation forced significant concessions from the examining countries that were reasonable within their existing laws. There were still some major issues to resolve.

5. Some examining countries agreed to reduced the refusal time limit to no more that 18 months, based on the report of a first office action. The U. S offered a 12 months refusal limit, pointing out at that time the average first office action was in six months. The textile, fashion and other design registration users liked the Draft Treaty features that combined to make their initial filing simpler and less costly. Another successful negotiation was tailoring the indirect filing application transmittal timing to meet the U.S. special needs for security review of all design patent applications. The time for forwarding the indirectly filed application was cut from three months to one month for all offices, except countries with national security review which had three months. In exceptional cases, with a formal request on each application a period of six months transmittal time could be obtained. This compromise was in the right direction, but it did not meet the U.S. needs.

6. There was a continued effort by countries with simple registration systems to press for accelerated examination for International Registration applications. This special treatment request was dropped, and replaced by a favorably received request for the option of accelerated examination available for all applicants.

7. Frequent reference was made to TRIPS art. 25(2) to facilitate the protection of textile designs. This concern seemed to be satisfied as the Draft Treaty concessions and clarifications evolved.

8. A concern expressed at this meeting was that the Draft Treaty should not create any conflict with other forms of intellectual property protection. While this issue had been addressed, a concern was expressed to clarify this principle.

9. A very important unresolved issue between the U.S. and the other countries was over the prior art effective date of an International Registration, in view of the *In re Hilmer* case law. If an International Registration application was based on a priority date under the Paris Convention, most countries wanted the prior art effect to be from the Convention date. The U.S. wanted the effective date for the International Registration to be when the International Registration application documents were received in the U.S., following established practice. In the U.S. the patent law prevented awarding the prior art effective date until an application was filed in the U.S. in acceptable form. This issue was another one that had to be

resolved at the Diplomatic Conference. The Sixth Session Draft Act contained an article X requested by the U.S. which allowed continue use of its *In re Hilmer* law. The overwhelming view of the delegates was to delete this provision. Even with this deletion, the Draft Treaty text was unclear on what would be the prior art effective date for an International Registration. The prior art effect issue remained a matter for national law in part, but the Draft Treaty would have to be clarified.
Progress was made in the Sixth Session, and the major issues were more sharply defined.

SEVENTH SESSION OF COMMITTEE OF EXPERTS ON THE DEVELOPMENT OF THE HAGUE AGREEMENT CONCERNING THE INTERNATIONAL DEPOSIT OF INDUSTRIAL DESIGNS, HELD AT WIPO, GENEVA, SWITZERLAND, Nov. 3-7, 1997

Summary

1. The seventh Meeting of Experts on the Hague Agreement revision had a renewed spirit in support of the Draft Treaty. This improvement came from the decisions made at the Sixth Session and a consultants meeting held by WIPO after the Sixth Session, with major interested countries, including the European Union, the U.S., Japan, and several non-governmental organizations, including UNICE.

2. The Draft Treaty prepared for this Meeting of Experts contained the major new compromise of 12 months to refuse International Registration acceptance. The compromise helped reduce the delay in protection in examining countries. In return, the Sixth Session Chapter II authorization to refuse filing for examining countries was deleted. A balance had been struck, to allow the negotiations to go forward, but many other issues remained. Some of these remaining issues were the independent voting status of the intergovernmental organizations, to vote along with its Member States, and the prior art effect of an International Registration. The vote and prior art effect issues were deemed too political for the experts to resolve, and they were deferred to the Diplomatic Conference.

3. A major WIPO proposal at the Seventh Session was to treat the Draft Treaty as a revision of the existing Hague Agreement, rejecting the alternative of a stand alone new treaty. The Draft Rules and selected Administrative Instructions were reviewed at this session. The discussion at this session made it clear that the owner of the International Registration selected its representative on matters before WIPO, and no special qualifications would exist for that representative. In the member countries it would be up to that office to determine their representation qualifications.

4. The roles of WIPO and the contracting party offices continued to be a major topic, with more discussion needed. These roles became clearer, with WIPO proposing it would have sole responsibility for the formality review, including whether a filing date could be obtained, and this proposal recognized the contracting party office role in examining offices to apply national law.

5. A likely benefit from the new Draft Rules will be the achievement of more uniform and simpler standards for International Registration use of drawings and photos. The Draft Rules required contracting parties to evaluate their drawings and photos procedures in light of procedures used by other contracting parties. Different practices and interpretations would be accommodated initially. After the treaty becomes operational, in the near future, more change toward uniformity will likely occur. One example of this type of change, noted by the U.S. delegation at this session, was the recent U.S. practice change that replaced the requirement for use of drawing shading with only a recommendation for use of shading. The U.S. delegation also referred to the U.S. recent practice change allowing use of color photos for design patents. These events evidenced the important role that international meetings on treaty development play in causing needed changes, even before a treaty is completed.

6. A new provision in the Seventh Session Draft Treaty mandated a minimum of six months from the International Registration date before the International Bureau could publish an International Registration, unless earlier publication was requested. This proposal was a part of a compromise to ensure that WIPO International Registration publication timing would include this delay, even if a country did not allow deferment.

7. The Seventh Session ended with the conclusion that no further Meetings of Experts were needed. The Draft Treaty, with a few adjustments made in a consultative fashion with interested parties, was ready for presentation to a Diplomatic Conference.

FINAL PREPARATION FOR THE DIPLOMATIC CONFERENCE

Summary

1. WIPO prepared the final set of documents for the Diplomatic Conference and had them reviewed by interested governments, intergovernmental organizations and non-governmental organizations. The changes made in the Seventh Session documents were to implement decisions made at that meeting. The introduction to the revised Draft Act had a summary of the main changes. These documents showed the prior text and changes made, making it easy to see the revisions. Some Draft

Act provisions were shifted to the Draft Rules, including details on procedures for refusal. These documents were submitted to the Diplomatic Conference.

DIPLOMATIC CONFERENCE FOR THE ADOPTION OF A NEW ACT OF THE HAGUE AGREEMENT CONCERNING THE INTERNATIONAL DEPOSIT OF INDUSTRIAL DESIGNS, GENEVA, JUNE 16 – JULY 6, 1999

Summary

1. The opening declaration by the government delegations, intergovernmental organizations and non-government organizations set the tone for the Diplomatic Conference. Several of the opening declarations are reproduced below. The spirit of cooperation and desire to successfully complete the treaty negotiations were evidenced. At the same time, as stated by the ABA representative, important issues remained to be resolved.

"*Opening of the Conference by the Director General of WIPO*

Mr. IDRIS (Director General of WIPO) welcomed the participants to the Diplomatic Conference. He noted that the Hague Agreement was the least known and the least used of the WIPO treaties for the registration of private industrial property rights, when compared to the Madrid Agreement and Protocol, and to the PCT. This was the reason why a Committee of Experts on the development of the Hague Agreement, which met between 1991 and 1997, had been established. The aim was to improve the system of international protection of industrial designs in order better to serve the needs of the creators and owners of industrial designs and to permit accession by countries which provided for an *ex officio* examination as to substance. He pointed out that the successful outcome of the Conference was dependent on the mutual understanding between groups of countries with different legal systems for the protection of industrial designs. He also expressed his confidence in the prevailing of a political goodwill to build a consensus over the issues relating to the status of intergovernmental organizations in the framework of the new Act. He said that both the Paris Convention and the TRIPS Agreement obliged Member States to provide a legal framework for the protection of designs, but that this answered the need only of those enterprises that could afford to file applications in all the countries where they were likely to be faced with counterfeiting. The Hague Agreement, however, made that protection more readily available. This was particularly important for small and medium-sized enterprises, not only in industrialized countries but also in developing countries." [WIPO Records of the Diplomatic Conference ... 388 (#1)(2002)].

Selected Opening Declarations

"Mr. LANDFERMANN (Germany), speaking in the name of the Member States of the European Community, congratulated the President on his election and expressed his confidence in the good progress of the Conference. The present Hague Agreement functioned without an examination prior to the grant of a design right, which had the advantage of suiting the specific character of a rapidly-evolving design industry. The present Hague Agreement was a fast and inexpensive system which offered effective protection. Under the new Act, two different systems would work together. This seemed to be the necessary compromise for the extension of the geographical scope of the Hague Agreement to many countries of the whole world including especially the emerging countries. The Member States of the European Community welcomed this compromise and proposed that the basic proposal be adopted without substantial changes. He expressed his special thanks to all those who had participated in the meetings of the Committee of Experts as well as to the staff of WIPO who had all put a great effort in preparing the revision of the Hague Agreement." [WIPO Records of the Diplomatic Conference ... 390 (#8) (2002)].

"Mr. HASHIMOTO (Japan) expressed his satisfaction concerning the election of the President and his conviction concerning the success of the Conference. The Japanese economy, which represented 16% of the world's economy, was still undergoing a difficult period, but it was expected to grow from the second half of this year. He believed that the easy obtaining of intellectual property rights in foreign countries played an important role in the growth of the world's economy. It was therefore desirable that everybody had access to the International Registration systems, in which WIPO played a central role. This was the reason why Japan had decided to join the Madrid Protocol for the International Registration of Marks. He informed the Conference that Japan's accession to the Madrid Protocol had already been approved by the Diet and that Japan would become a party to the Madrid Protocol next January. In the area of industrial designs, he was convinced that Japan's accession to the Hague Agreement would have a beneficial impact in the revitalization of the world's economy. Indeed, the number of applications submitted to Japan each year represented a substantial part of design applications in the world. However, in order for Japan to be able to join the Hague Agreement, some flexibility was required from the countries which had a different system. Japan encountered two main problem areas in the new Draft Act of the Hague Agreement. The first one, which was shared by other Asian countries, was a language problem. If Japanese residents used the Hague Agreement to seek protection in Japan, the Japanese Patent Office would receive some 40,000 applications in English each year. This would render difficult the work of that office. He therefore hoped that an optional provision, enabling the prohibition of self-designation, would be introduced in the new Act. The second problem area was common to those countries whose offices conducted a substantive examination. It related to certain provisions such as those concerning fees, the effects of the International Registrations, refusal, publication six months after registration, and the transfer of the design rights. Finally, he stated that an online design application system was soon going to be available in Japan and that publication via the Internet was already operating. He believed that the adoption of the new Act of the Hague Agreement would increase the efficiency of the International Registration system and would bring about a global expansion in the use of electronic information on design rights." [WIPO Records of the Diplomatic Conference … 391 (#19)(2002)].

"Mr. MARCHANT (United Kingdom) congratulated the President on his election and thanked the Secretariat for organising the Conference. The driving force leading to the Diplomatic Conference had been the recognition that the present Hague system was unsuitable for those countries such as the United Kingdom which operated an examination system. The extension of the geographical coverage of the Hague system was very important. However, the system that would be put in place would also have to be user-friendly. The UK Delegation would seek an agreement that balanced the need to attract wide participation with the need to provide a customer-friendly system. The UK had first-class designers showing considerable interest in protecting their designs abroad, who were well aware of the potential usefulness of the Hague system. However, about two thirds of the designs registered in the United Kingdom each year were of foreign origin, so that if the United Kingdom ratified a new Act of the Hague Agreement, the greatest benefit in the UK could go to designers from abroad. This is a challenge the United Kingdom is willing to face. Mr. Marchant stated that the handling of designs in the UK was being altered and that the law was being amended to comply with the European Community directive on the legal protection of designs. Furthermore, UK registered designs have begun to be published on the Internet, so that they can be inspected and searched without charge. In addition, registration procedures have also been reviewed, which has resulted in a speedier service for customers. He emphasized that rapid access to rights was increasingly important in a fast-moving and highly competitive environment. He finally expressed his confidence in the successful outcome of the Conference and assured it that the United Kingdom Delegation would approach the negotiations in a most constructive and positive way." [WIPO Records of the Diplomatic Conference … 391-392 (#20)(2002)].

"Mr. BOUHNIK (Algeria) congratulated the President on his election and thanked the World Intellectual Property Organization for the excellent work it had accomplished. He stated that his country was perfectly aware that, confronted with the new worldwide context marked by the profound changes that were taking place in the economic environment, the system for protection of industrial property, including designs, needed to be continuously adapted in order to meet those challenges. That was why his country was enthusiastically supporting the proposed new Act of the Hague Agreement. It was for that same reason that Algeria was envisaging, in particular, attending the forthcoming Diplomatic Conference on patent law. The Hague Agreement made it possible to provide uniform and effective protection in all the States bound by that Agreement and contributed to achieving national objectives. However, the question of the real chances for a developing country such as Algeria to draw benefit from the system remained relevant. A start had been made on giving a reply to that question at the orientation meeting on the Hague system held the day before the Conference. The remainder would certainly depend on the manner in which the participants at the Conference were willing to show flexibility in order to achieve compromises. Mr. Bouhnik observed that, with goodwill and the spirit of international solidarity, the fundamental issues contained in the Basic Proposal could be resolved. It was with that hope that the Algerian Delegation was participating in the Diplomatic Conference and it would spare no effort to ensure its success." [WIPO Records of the Diplomatic Conference ... 392 (#22)(2002)].

"Ms. SUMEGHY (Hungary) congratulated the President on his election and expressed, on behalf of her government, her gratitude to the Director General of WIPO for the invitation to such an important conference. She pointed out that Hungary had been a party to the Hague Agreement Concerning the International Deposit of Industrial Design since 1985 and that the system established by the Agreement had increased the interest for foreign applicants in seeking protection in her country. The statistical data of the last five years showed, on the one hand, a continuous increase in the number of International Registrations designating Hungary, and, on the other, a rather stagnant number of applications filed through the national route. Hungary participated in all sessions of the Committee of Experts and was certainly in favor of the initiative to establish a new Act with the purpose of attracting new countries, as well as to improve the current system. She pointed out that, although the new Act does not attempt to harmonize the substantive provisions of national legislations on industrial designs, it will certainly require some adjustment at national level, for instance, as far as the term of protection in concerned. In this context, she mentioned that in the near future Hungary will start preparing a draft for a new law on the protection of industrial designs in line with the relevant European Community directive and this will represent a valuable opportunity to take into account also the provisions of the new Act. In light of the above, Ms. Sumeghy concluded by stating that the Hungarian Delegation fully supported the adoption of the new Act, and intended to ratify it in the not too distant future." [WIPO Records of the Diplomatic Conference ... 392-393 (#23)(2002)].

"Mr. KIM (Republic of Korea) thanked the President and congratulated him on his election, confident that his leadership abilities and dedication would ensure the success of the Conference. He also expressed his sincere appreciation to the members of the International Bureau who prepared the draft new Act of the Hague Agreement, and made arrangements for the Diplomatic Conference. While the Republic of Korea had participated in all seven sessions of the Committee of Experts and therefore was in principle in favor of the adoption of the new Act of the Hague Agreement, careful thought should be given to the needs of countries having an examination system and this should be reflected in the new Act. In this regard, he expressed his fear that the limit for the determination of individual fees might be fixed at too low a level. Furthermore, he pointed out that the Republic of Korea strongly supported the proposal of Japan concerning the introduction in the new Act of a provision prohibiting nationals of a particular country from designating that country in an international application, since, otherwise, the Korean Office would receive a very substantial number of applications in English and French also from nationals of the Republic of Korea and this might entail a serious risk of delaying the granting of protection." [WIPO Records of the Diplomatic Conference … 393 (#24)(2002)].

"Mr. DICKINSON (United States of America) congratulated the President on his election and the International Bureau of WIPO for the excellent work carried out during the seven sessions of the Committee of Experts and in connection with the Diplomatic Conference. He recalled the importance of designs as a factor of commercial success not only, as in the past, for luxury products, but also for mass-market goods. The growing necessity to devise new designs to differentiate goods produced by different undertakings entailed a significant increase in the number of industrial designers employed in the United States of America (50% more over the past five years). Furthermore, contrary to what happened in the past, it was no longer possible to limit the circulation of products, and therefore this potentially unlimited dissemination (also through the Internet) of industrial designs gave rise to an increasing risk of counterfeiting. This, in turn, rendered it even more vital to be able to obtain effective protection for industrial designs also at international level, possibly through a new system which constituted an improvement with respect to the existing one and was compatible with the many systems existing throughout the world, including that of the United States of America. Mr. Dickinson concluded by recalling the great interest which the American industry has shown in a prompt adoption of a new Act." [WIPO Records of the Diplomatic Conference … 395 (#35) (2002)].

"Mr. FRYER (ABA) congratulated the President on his election. It was his privilege to represent the American Bar Association which supported completion of the revised treaty to the extent that it was compatible with current United States design patent law. ABA did not however support a revised treaty that would require significant changes in U.S. law. By way of example, he expressed doubts that a change to remove the *In re Hilmer* law, which denies prior art effect for foreign-filed Paris Convention applications back to their priority dates, would be made in the near future in the United States. He noted that such a change would be a matter of substantive law, which was not the purpose of the revised Act. Mr. Fryer stressed that the current situation in the United States on the *In re Hilmer* issue was essentially the same as in 1970, when the Washington Diplomatic Conference on the Patent Cooperation Treaty was held. On that occasion, the participating States had recognized the need to provide a reservation for the United States on the issue of prior art effect, to allow retention of the *In re Hilmer* law, and he urged the members of this Diplomatic Conference to adopt the same type of reservation. He then indicated that the six-month maximum examination period for national offices provided for in the 1960 Act of the Hague Agreement was unrealistic and that the task of the Conference was to adopt realistic provisions for a global design application filing system." [WIPO Records of the Diplomatic Conference ... 397 (#43)(2002)].

"Mr. ADAMS (ICSID) thanked the President and congratulated him on his election. ICSID had participated in all seven Committees of Experts and strongly supported the principle of widening the geographical scope of the Hague Agreement since with the increasing globalization of trade, an effective system of international protection of industrial designs plays a pivotal role. The vast majority of designers were small or medium-sized enterprises; if any new Act was to be attractive to them, it needed to be as simple, speedy and cost-effective as possible. Furthermore, referring to the encouraging steps which had been taken towards the unification of design legislation in the European Union, Mr. Adams stressed that the way in which design was practiced at the end of the twentieth century had diverged considerably from the way in which the protection was often granted by national laws; he therefore suggested that any government considering updating its laws could follow the approach undertaken by the European Union. He finally indicated that his organization attached the greatest importance to the successful outcome of the Conference." [WIPO Records of the Diplomatic Conference ... 398 (#48)(2002)].

2. The experts had prepared a very thorough Draft Act and Rules for the Diplomatic Conference. Many issues were settled at the Meetings of Experts and approved at the Diplomatic Conference without changes. This summary will not discuss in detail topics settled at the Meetings of Experts and adopted without change at the Diplomatic Conference, unless they related to new developments occurring at the Diplomatic Conference. There were many important issues that had to be resolved at the Diplomatic Conference.

The Diplomatic Conference followed a procedure of submitting proposed Draft Act and Rule changes in written form. Many delegations submitted proposed amendments. The largest number of proposal documents (8) was submitted by the U.S., indicating that a careful review was made and a desire on the part of the U.S. to make the Geneva Act compatible with the U.S. design patent system. While the U.S. was very involved in the Diplomatic Conference, other countries, intergovernmental organizations and non-governmental organizations were equally active in developing suitable compromises.

The most challenging issues were the European Union (EU) proposal on the right to vote separately, in addition to its Member States in the administration of the Geneva Act, and the requirement that an International Registration have the same prior art effect as a national application. In understanding the negotiations that took place, all issues have to be considered. The U.S. and other examining countries were particularly concerned with the time limits and effect of the International Registration refusal provisions. The non-examination countries and intergovernmental organizations were convinced that negotiations on the Draft Act had to be used to make the examination systems more effective. The tension between the world's two major types of design registration systems was at the center of most of the issues discussed.

3. The EU was separately represented at the diplomatic conference as one of the intergovernmental organizations. Its role in the Meetings of Experts and the Diplomatic Conference was very important. At various times the States in the EU spoke with individual points, but on some issues like the separate vote of intergovernmental organizations, the EU spoke with one voice.

After the Geneva Act approval, the Hague Union assembly would vote on changes to the Act and other matters. The issue of an intergovernmental organization's right to vote separately had to be resolved. The resolution of this issue came at the very end of the Diplomatic Conference, with the agreement that an intergovernmental organization could vote if none of its States voted at the same time. When any State in the intergovernmental organization voted on a matter, the intergovernmental organization could not vote. This voting procedure issue had been raised in connection with several international negotiations with the EU, and it was the Geneva Act that took the issue to a successful conclusion.

4. The debate on the prior art effect of an International Registration was primarily between the U.S. on one side and most of the other delegations on the other side. There was a long history on international negotiations in several forums to change the U.S. utility patent law, based on the *In re Hilmer* case, to provide prior art effect for an issued U.S. utility patent from the Paris Convention foreign filing date for a U.S. utility patent. Since U.S. design patent law is governed by the general patent law, the same interpretation applied to design patents.

At the beginning of this Diplomatic Conference the representative of the American Bar Association presented a statement [*see* Selected Opening Declarations above], that reviewed the history of the Patent Cooperation Treaty (PCT) Diplomatic Conference on this issue. He indicated that U.S. law on the prior art effect of a patent was not changed as a result of the PCT negotiations. Recent publications and discussions had shown some interest in considering a change in the this law.

Several very important compromises were made that allowed the U.S. to interface with the Geneva Act to retain its basic design patent system without major change. This good faith effort helped the U.S. accept the prior art effect on the International Registration date. This date was when the International Bureau received the International Registration application directly, or the dated filed indirectly in a contracting party, if it is received by the International Bureau within the time limit.

The Geneva Act has a general provision on International Registration effect. The prior art effect result is one of the effects from this provision. It gives the international application, when registered, the same effect as a national application on that date.

5. There were several aspects of International Registration effect refusal by an examining office that needed to be negotiated at the Diplomatic Conference. The experts proposal was accepted that a refusal was satisfactorily made if it presented what was known by the national office at that time. Later, other issues could be raised. This type of refusal response would be like a U.S. first office action, for example.

The U.S. submitted a proposal at the Diplomatic Conference that the maximum refusal period should be 18 months, but it was not advanced forcefully. The Meeting of Experts proposed a 12-month period that was accepted at the Diplomatic Conference. It was a major compromise for the U.S. and consistent with making examining systems more effective.

Another U.S. concern about the refusal procedure in the Draft Act was the automatic initiation of effect, creating rights, if no refusal was received. This failure to respond, due to administrative error was a problem for large examining offices. A provision was included in the Geneva Act that allowed for administrative adjustment to avoid that situation.

6. The experts had accepted a minimum 15-year International Registration protection term. This provision was accepted at the Diplomatic Conference. It was a significant increase for some States. There was very little discussion at the Diplomatic Conference on reducing this term.

7. There was significant discussion on the unique U.S. patent procedure feature of requiring a national security review before transmitting a design application out of the country. The Draft Act proposed a three-month period for offices with a national security review to complete the review and deliver the indirectly filed International Registration to the International Bureau. If a delay was expected on an application, the office had to inform the Bureau on that case and another three months was added to the limit, for a total of six months. The Diplomatic Conference accepted the U.S. proposal for a general six-month period, eliminating the separate correspondence. The U.S. national security review had its own cycle and working with it was considered best for efficient processing.

8. The Meetings of Experts had no solution for the U.S. concern on the effect of the International Bureau recording of ownership changes. In the U.S. this recording has significant legal effect on ownership rights. The Draft Act gave the Bureau recording of ownership legal effect in all contracting parties. The U.S. proposal was accepted that the legal effect of the Bureau recording would take place only when the document was received by the PTO. This revision avoided a major conflict that could have kept the U.S. from becoming a member of the Geneva Act. A change in U.S. law, to follow the Draft Act proposal, would have involved all patents, broadening the debate.

9. The Meetings of Experts had worked out a procedure for examining offices to receive copies of deferred International Registrations if an office made a general request for them and they were designated in the International Registration. The Diplomatic Conference recognized the importance of the deferred International Registration as possible prior art in examining and non-examining systems. The Diplomatic Conference revised the procedure and allowed any designated contracting party in the International Registration to receive the deferred registration copy.

In hindsight, it is likely that deferred International Registrations will be important to all contracting parties in the novelty evaluation that occurs in the office examination, post-grant opposition, or infringement and validity litigation. This revision was an important step.

10. One interesting change in the Draft Act may have been a sign of future changes in The Geneva Act. The EU successfully urged approval of a change that allowed intergovernmental organizations to set an individual designation fee. The Draft Act had stated that only examining offices could use the individual designation fee to cover administrative costs, instead of the standard designation fee that the non-examining offices had to use. After appropriate revision of the Geneva Act, the EU request may be a step other non-examination offices will want to take, to keep up with increasing local costs.

11. Discussions in the Meetings of Experts made it clear that national publication of International Registrations was a major concern for all design registration offices. Some offices wanted to publish the International Registration to make it accessible in the local language. Their problem, under the Draft Act, was that no fee could be charged for the publication of an International Registration under most circumstances. The International Bureau publication was mandated as sufficient.

There was a basis for agreement on how to revised the Draft Act. If the examination resulted in a change in the design, the changed International Registration could be published and a fee charged. The concept behind this agreement was that a change in the International Registration that effected the design right has occurred. Merely translating the International Registration would not justify a re-publication fee.

12. The Draft Act had a requirement that each office would report to the International Bureau when an International Registration was invalidated, so this information could be put in the Bureau records and published. Many delegations pointed out that national offices do not have all this information. The resolution of this issue was to require reporting only the information that the office had available. The status quo prevailed.

13. The successful completion of the treaty was followed by Mr. Churchod's concluding remarks in behalf of Director General, Dr. Kamil Idris:

"Mr. CURCHOD (WIPO) recalled that the Director General, Dr. Kamil Idris, who was unfortunately absent from Geneva, had said in his opening address that the Diplomatic Conference was the result of more than eight years of preparatory work by the Committee of Experts and that it was necessary that those countries with an examining office and those whose office did not carry out an examination of novelty should show a mutual understanding to ensure that the Conference was a success. That spirit of cooperation had obviously reigned. Indeed, it had been shown by all the participants. The new Geneva Act of the Hague Agreement, that had just been adopted, achieved a delicate and subtle balance. It took into account the specific interests of the potential Contracting Parties having an examining office whilst at the same time maintaining the basic simplicity of the Hague system. The latter point was most important for users, as was the prospect of extending the geographical scope of the Hague system. Furthermore, it was particularly welcome that the issue of the status of intergovernmental organizations within the Hague Union Assembly had been amicably settled, taking up a minimum of time in the formal meetings. That had enabled the Main Committees to focus on matters directly related to the international protection of industrial designs. However, the biggest reason for satisfaction at the International Bureau was to be found in the fact that the Geneva Act of the Hague Agreement had been adopted by consensus. That fact was most gratifying both for the Secretariat and for WIPO as a whole. The wishes that had now to be expressed were that the greatest possible number of delegations should sign the Geneva Act on July 6 or within the year that followed, and that those signatures should be followed by ratification in the nearest possible future. Mr. Curchod finally echoed the congratulations that had been paid to all the Presidents who had officiated during the Diplomatic Conference. He invited all the delegations to be present on Tuesday, July 6, in the main building of WIPO for the signature of the Geneva Act of the Hague Agreement Concerning the International Registration of Industrial Designs and of the Final Act of the Diplomatic Conference." [WIPO Records of the Diplomatic Conference ... 523 (#1139)(2002)]

14. The next stage of the diplomatic conference was the signing of the new Geneva Act by governments and intergovernmental organizations. This step indicated their commitment to work toward implementation of the Act in their countries and

intergovernmental organizations. On July 6, 1999, the following delegations signed the Geneva Act of the Hague Agreement Concerning the International Registration of Industrial Designs, adopted at the Diplomatic Conference on July 2, 1999. Algeria, Belgium, Bulgaria, Chad, Croatia, Cuba, Denmark, Estonia, France, Greece, Hungary, Italy, Latvia, Netherlands, Portugal, Republic of Moldova, Romania, Russian Federation, Slovenia, Spain, Sudan, Switzerland, United Kingdom and United States of America (24) [WIPO Records of the Diplomatic Conference … 194 (2002)].

CHAPTER 2

Operation – Introduction to the Geneva Act (1999)

The central tools used to introduce the 1999 Act operation are Figures 1 to 3 in this chapter. Figure 1 is a diagram of the Act's operation, from the perspective of a design owner entitled to file in a contracting party that registers designs without a novelty examination (non-examination office). Figure 2 is a diagram of the Act's operation from a design owner's perspective entitled to file in a contracting party that has a novelty examination office (examining office). Figure 3 illustrates the timing aspects of the 1999 Act. Taken together these figures help explain the basic concepts, terms and timing of the 1999 Act.

In the figures, for the convenience of the reader, reference numbers in square brackets have been inserted to mark the location in a figure that is being described.

The Topics chapter is the reference source for further study of the concepts, terms and timing relationships presented in these figures. For the convenience of the reader, in this chapter references to related 1999 Act articles, Rules, Administrative Instructions and Meetings of Experts documents have been omitted.

National Design Protection Systems

As a starting point in reviewing the 1999 Act, it is helpful to provide a general background on the two main types of national design registration systems and their relationship to other forms of IP protection. Most countries have more than one system for protecting intellectual property. The patent system (utility patent in the U.S.) protects primarily product functionality and structure – how a product works or is made. Industrial design systems, either design patent or design registration, protect the product appearance. In this book the terms industrial design registration or design registration, include the design patent. The term patent is used in this book to refer only to the patent protection for functional and structural features, unless otherwise indicated.

There is an overlap between the patent and industrial design registration, due to the fact that the appearance may include some of the functional product features. National laws have to deal with this overlap, and the approaches vary. For most systems industrial designs cannot be registered if the design is primarily functional.

There are two fundamentally different national industrial registration systems, the non-examination and the examining types. The non-examination system does not review the novelty of a design before registration. The examination system requires that standards of novelty and, in some cases, non-obviousness, be met before registration.

Figure 1 shows a national non-examination system (at point [1]), in a contracting party X, which may be a country or a regional system administered by a intergovernmental organization. The non-examination system has a rapid registration process. An application is filed showing the design and providing required information. If the formal review finds the application has the required contents, protection is granted. Terms of protection vary, with some up to 25 years. Usually renewals must be obtained to continue protection for the maximum period. If there is a dispute as to whether novelty standards under the law are met, these matters come up after the grant, either in an administrative proceeding or in court proceedings. The national office checks the application formal aspects, handles the record keeping, industrial design registration, publication and renewals. Usually there is a separate industrial design law not related to the patent law that sets up the non-examination system.

A national examining registration system is shown in Figure 2, in the lower area, at location [1]. In this example the system described is the U.S. design patent system. Applications for registration are filed, and after the formalities are checked, the design is examined for novelty and non-obviousness. If the standards are met, the application is allowed and rights are granted that begin at the time of the grant for a period of 14 years.

Figure 2 also illustrates at location [3] a unique feature of the U.S. system that requires national security review of each design registration before any foreign filing can occur. A foreign application can be filed after the security review and a license to file is obtained, and the filing under the Paris Convention for priority must not be more than six months from the U.S. application filing date. It is apparent that design foreign filing is a busy job if all the requirements are to be met in a short time. In the Figure 2 example, the International Registration has a priority date of a prior filed U.S. application, as it was filed within the six-month Paris Convention period.

Legal Relationship of the National Industrial Design Registration Systems and the 1999 Act

A key concept to understand and one that takes some time to comprehend fully, is that the 1999 Act sets up an international protection system. While it incorporates national industrial protection law, its legal status, operation and administration is under an international treaty. Countries will approach the interface with treaties in different ways. The 1999 Act separate treaty protection system using the International Registration (IR) is illustrated in the upper portions of Figures 1 and 2, at location [2]. The International Bureau (IB) at the World Intellectual Property Organization (WIPO) administers the treaty, performing the formal review and

Figure 1. Non-Examining System Example – Use of 1999 Geneva Act – Design Owner's Perspective

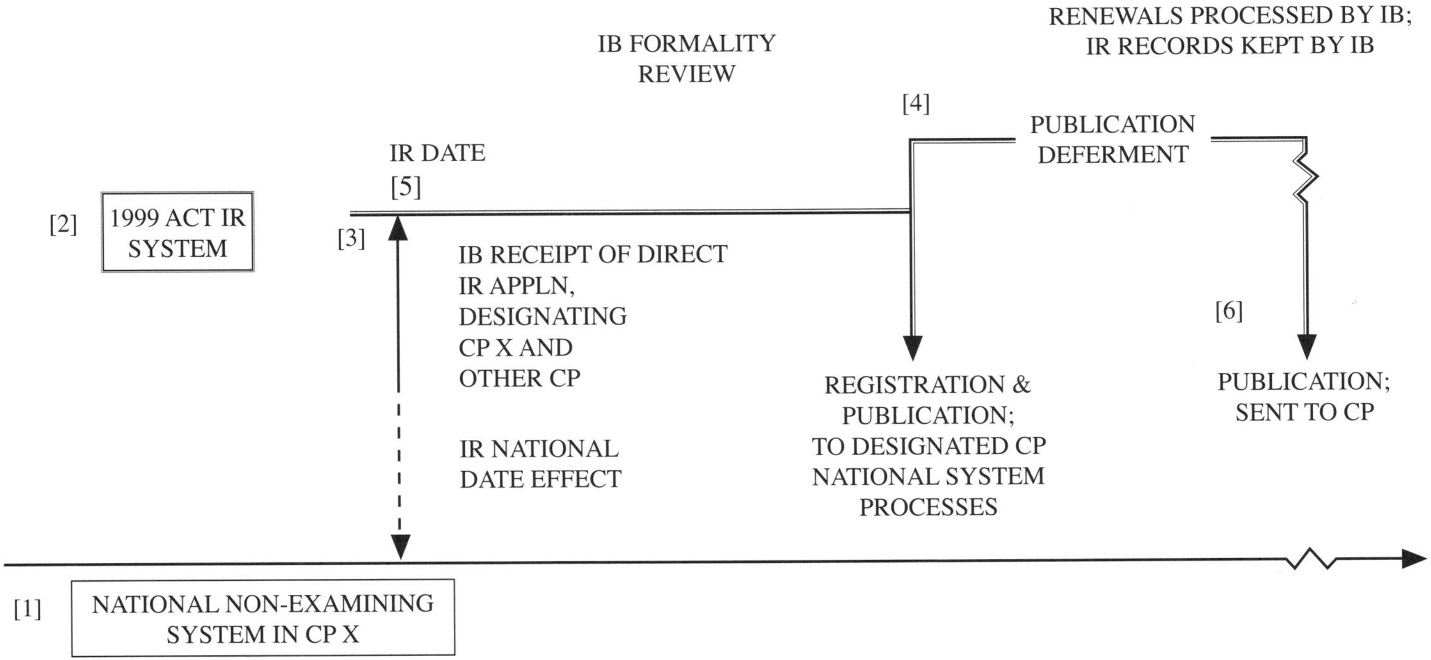

LEGEND:

APPLN – Application
CP – Contracting Party
IB – International Bureau of WIPO
IR – International Registration
1999 Geneva Act, Hague Agreement Concerning the International Registration of Industrial Designs

Figure 2. Examining System Example – Use of 1999 Geneva Act – Design Owner's Perspective

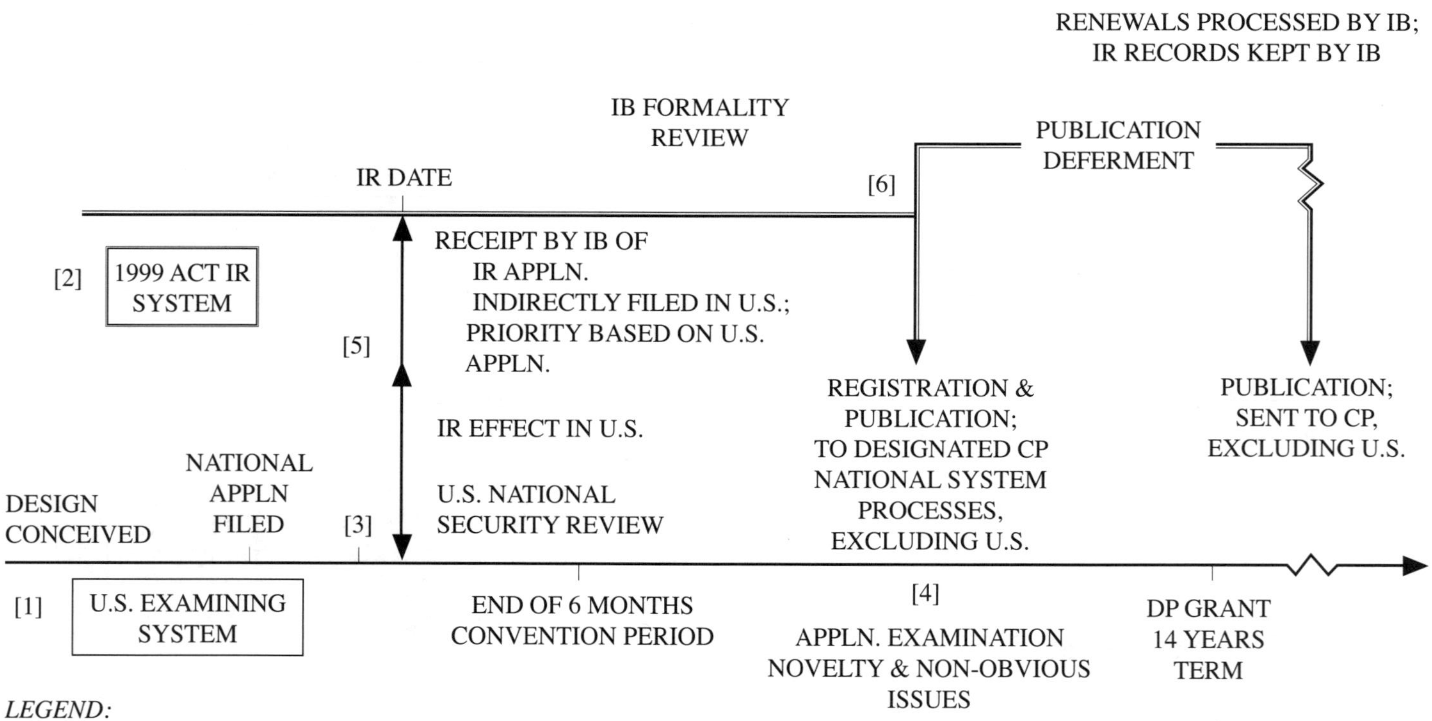

LEGEND:
APPLN – Application
CP – Contracting Party
IB – International Bureau of WIPO
IR – International Registration
1999 Geneva Act, Hague Agreement Concerning the International Registration of Industrial Designs

keeping the records and handling the renewals. A country or intergovernmental organization that joins the Hague Union under the 1999 Act is called a contracting party (CP).

General 1999 Act Operation for a Non-Examining Contracting Party

The 1999 Act interface with a contracting party having a non-examining office is illustrated in Figure 1. There are several important features to recognize. In this example the International Registration application is sent directly to the IB, as indicated at location [3], after appropriate security approval, for the U.S. applicant. The International Registration date, indicated at location [5], is when the International Bureau receives the complete IR application for the designated contracting parties.

The International Registration application will designate the contracting parties where the protection is sought. After the International Bureau formality review and, if the application is complete, the registration is published, at point [4] in Figure 1, and this publication has the effect of a filed national application in the designated parties, as of the International Registration date, shown in Figure 1 location [5]. Another option exists instead of publication. The publication may be deferred at as shown in location [4] in Figure 1, depending on which countries are designated. These features are discussed in more detail below. The effect of the International Registration when accepted, whether published or deferred, is the same, providing a filing date in the designated countries on the International Registration date, and having the same legal effect as a national application in the contracting party.

The designated national offices begin processing the International Registration, the same as a national application after deferment ends, at point [6] in Figure 1. In the case of a non-examining system, shown in Figure 1, lower level at point [1], its office will follow the same procedures for the International Registration as for the national application, with some time limit constraints (*see* Figure 3 for more details on The 1999 Act timing relationships).

It is important to review the main features of a typical non-examination national system shown in Figure 1. Much of the discussion on how to revise the Hague Agreement was based on the simple operation of a non-examining system. The challenge was to adopt a revised treaty that would work for both non-examining and examining systems. In addition, the non-examining system countries wanted a more efficient and quicker review in examining systems. In this way an attractive global treaty for protection of designs could be achieved.

A national application, especially in a non-examination contracting party, may offer the opportunity to include up to 100 designs in the registration application. This procedure usually reduces application fee costs, as filing and renewal of additional designs in the application cost a lot less. The registration is for all the designs in the application. The limit on using the multiple designs feature in a national system generally is governed by the International Classification, under the Locarno Agreement, requiring that all designs in an application be within the same international class.

In some examining national systems a unity of design concept is applied, meaning only one design concept can be examined in an application. If more than one design concept is in an application, the national procedures usually allow the separate designs to be processed as separate applications, retaining the original filing date. Regular filing and other fees must be paid for each of the separate design applications.

The International Bureau's role in International Registration administration is to handle the receipt of applications, record keeping and publication of the International Registration. All fees are paid to the Bureau in Swiss currency. Copies of published International Registrations are sent to each designated contracting party. When there is deferment, contracting parties designated in an application can request copies of the deferred registration.

In a non-examining contracting party the effective date for rights granted to bring an infringement action will be the same as for a corresponding national registrations. Some systems delay the grant of rights until certain procedures are completed. Other systems provide that the rights become effective at a certain time and go back to the filing date of the application. A significant number of national design registration systems make the filing date the effective date for rights granted upon registration.

The International Registration deferment option follows in the footsteps of many national systems that have used this feature, and the earlier 1934 and 1960 Hague Agreement Acts that had deferment. There is a historical need for design registration deferment in the opinion of some industries, like textiles, fashion and automobiles, to prevent use of a published design registration as early information to competitors on products under development.

The 1999 Act allowed up to a 30-month deferment from the International Registration date, at the option of the contracting party. During that time no information on the registration is published by the International Bureau. It is a confidential document and access is limited during the deferment only on a confidential basis for certain situations. The presence of a deferred International Registration which may have enforceable rights under national laws creates a situation that has to be addressed in each contracting party national system. The primary concern is prior art effect, so that provision is made for each national office that is designated in a deferred International Registration to obtain, if requested, a copy of all deferred International Registrations for its records.

Once the deferment period ends, the International Bureau publishes the International Registration, as shown at location [6] in Figure 1. The designated contracting parties start their review and a refusal must be made in six months for non-examining contracting parties. After the refusal the review continues until completed, following the same procedure used for national applications. The date when rights are granted depends on national laws, and it is likely to be the same date as for corresponding national applications. If there is no refusal within that six-month period, the International Registration automatically has effect in the designated contracting party as of the International Registration date.

A very special procedure for International Registration of any two-dimensional design was set up to assist the textile industry. If deferment is requested for these designs, a specimen of the product with the design, like a fabric sample, can to be filed with the application to obtain a filing date. The publication fee, usually paid at the time the application is filed, need not be paid until just before the deferment period ends. A design owner can renounce the protection of one or more designs in the International Registration, or eliminate protection in selected contracting parties for one or more designs, before publication.

There may be significant cost saving from the simplified filing procedures for two-dimensional designs, coupled with the features of multiple designs in one International Registration. The procedures allow a business to focus on continuing protection on more valuable designs. These features are very helpful to the textile industry. Other industries receive the same benefits from deferment and the multiple design application features. The textile industry has the added advantage of using specimens and the reduced initial filing costs.

The International Registration protection term is a minimum of 15 years from the International Registration date. It will be as long as the national design registration. The International Bureau handles the five year renewals.

General 1999 Act Operation for an Examining Contracting Party

Figure 2 shows the U.S. design patent system, as an example of how an examining system would interface with the 1999 Act. The U.S. is not a member of the 1999 Act at this time. The lower part of Figure 2, at [1] is the U.S. national system.

If the U.S. was a contracting party under the 1999 Act, a U.S. design owner would have several options to initiate foreign protection. The Figure 1 approach of direct national filing could be used. Another choice would be to indirectly file an International Registration in the U.S. Patent and Trademark Office (PTO) and designate contracting parties for protection. The PTO would send the International Registration application to the International Bureau, as indicated at location [5], after conducting the national security review indicated at location [3].

If the U.S. would allowed self designation, i.e. filing in the U.S. through the International Registration, where the U.S. is the applicant's contacting party, all the desired filing would be completed in one step. Self designation is a practical feature that should be available to U.S. design owners. This International Registration has a priority date of the earlier filed U.S. filed application, as it was filed under the Paris Convention within the six months from the U.S. application filing date.

The national application moved through its examining office procedure, where the novelty and obviousness were examined at location [4]. After approval, the design patent was granted for a term of 14 years. There was no protection from infringement prior to the grant of the design patent.

The U.S. design owner can include multiple designs in the International Registration. Several pending U.S. design patent applications can be in the International Registration application, if they are in the same international class, relying on the priority dates of the prior filed U.S. applications. There are significant cost savings from using the multiple design feature. In addition, combining multiple designs with deferment has additional potential cost saving benefits.

It is likely that the International Registrations designating the U.S. would not be deferred, as the U.S. indicated at the preparatory meetings and Diplomatic Conference that deferral is not an acceptable practice in the U.S. Under the 1999 Act procedures, the contracting party with the shortest allowed deferment period controls the deferment for all designations in an International Registration application. The U.S. designated International Registration would be sent to the PTO immediately upon publication, as shown at location [6], and processing would begin, following the U.S. design patent law and procedures. The International Registration will be published on the Internet, through the WIPO website on the day of registration and the national offices will have the IR information the day it is published to the world.

The order of examination in the PTO would be the same for all applications, International Registrations and national applications, essential in the order filed. The only commitment made by the U.S. was to have a first office action within 12 months. Therefore, the U.S. would likely make a declaration to have the additional six months, and the refusal period would be a maximum of 12 months. At the time of the Diplomatic Conference the PTO was mailing first office actions on all design patent applications well before 12 months from the U.S. filing date. There should be no special treatment for International Registrations. Only a first office action is needed within the time limit, and new grounds for rejection can be raised after that time, such as rejection on new prior art.

If the U.S. design owner had used the IR to obtain protection in the U.S. as well as other contracting parties, the procedure in the review of the IR would begin at the time of publication. The IR would go through the same procedures set up for national design patents.

The effective date of the U.S. International Registration rights in the U.S. will be set up by a declaration to the WIPO Director General, probably beginning at a time of or shortly after allowance of the International Registration. The term will be the same as under U.S. design patent law. It is likely that the U.S. would require full payment in the individual designation fee for examination, and have the publication fee and full term protection cost in a separate fee, the current design patent

Figure 3. Time Features of the 1999 Geneva Act – Design Owner's Perspective

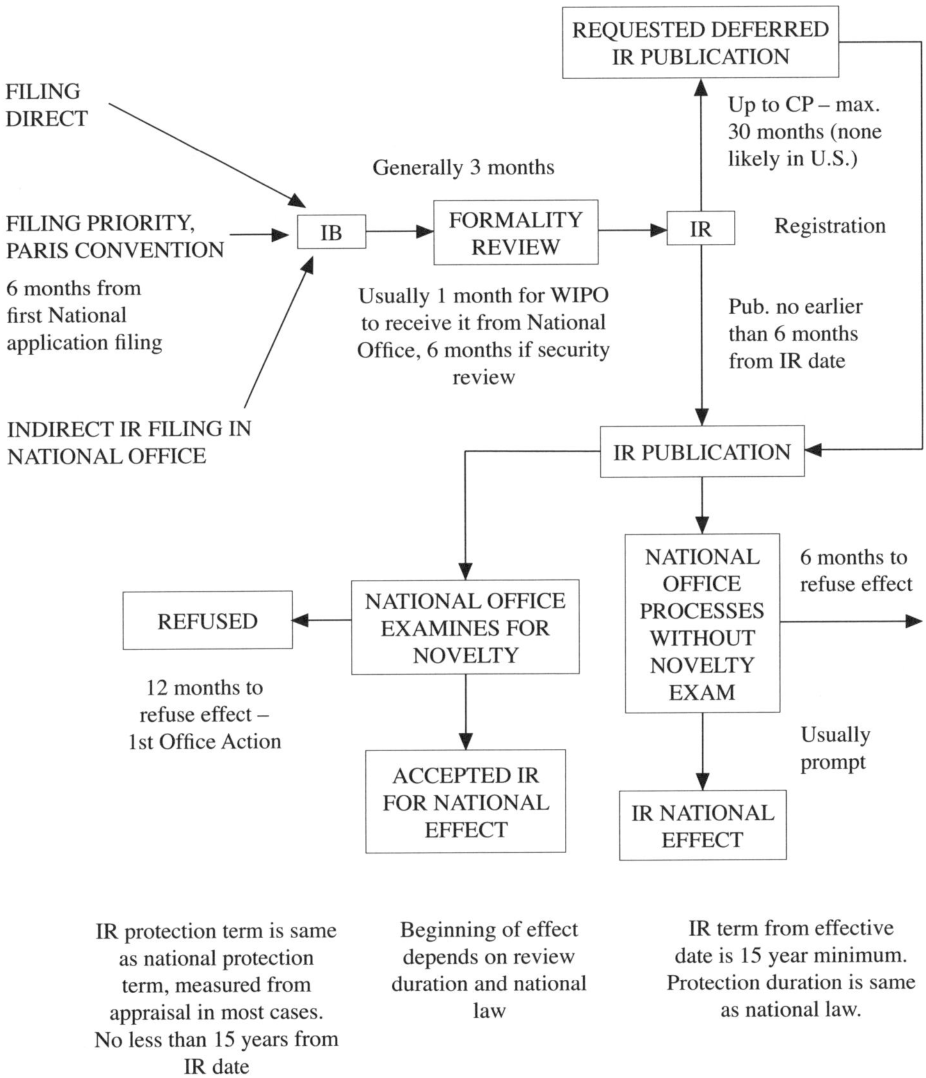

LEGEND:
IB – International Bureau of WIPO
IR – International Registration (design)
CP – Contracting Party
1999 Geneva Act of the Hague Agreement Concerning the International Registration of Industrial Designs

procedure. This arrangement would treat foreign applications and U.S. applications the same. There is no renewal of a design patent protection term in the U.S.

There is a Geneva Act (1999) principle that must be applied to determine where a design owner designates protection. It is a starting point for developing protection strategy. Assuming the U.S. is a member of the 1999 Act, in the case of the U.S. design owner discussed above for Figures 1 and 2, the entitlement to use the 1999 Act comes from being a national, or domicile, or having an appropriate business establishment in the U.S., or being a habitual resident of the U.S. An International Registration can only designate protection in contracting parties that are members of the same Hague Agreement Act, as the applicant's contracting party. The U.S. is not a member of any of the other Hague Agreement Acts. Therefore, a U.S. design owner filing based on the U.S. entitlement would be able to designate only the 1999 Act contracting parties for protection. If a contracting party had more than one common treaty relationship with the originating contracting party, the latest one would apply. In practical terms, the evolution of the Hague Agreement system is to the 1999 Act, and the benefits of global protection will be based on that Act.

Non-Examination and Examination Contracting Parties Time Features of the 1999 Act

With the previously described non-examination and examination systems shown in Figures 1 and 2 in mind, it is useful to review the time features for these systems as applied to the 1999 Act. Figure 3 illustrates the basic sequences and timing.

The starting point is the International Registration filing. The International Registration date is when a complete application is received by the International Bureau. The indirectly filed International Registration date is the national office filing date, if the International Registration application is forward to the International Bureau on time and it is complete. There are mandatory content items to receive before a filing date will be awarded. If there are some items missing, it may be that the filing date will not be awarded until after receipt of those items. For examining contracting parties there may be additional content items required. While there may be a priority benefit from an prior filed application under the Paris Convention for an International Registration, it does not effect the International Registration date.

The International Registration application formality review usually takes no more than three months. Under the 1999 Act, publication will occur no earlier than six months from the International Registration date, unless earlier publication is requested. The designated offices will start their processing upon International Registration publication, with the International Registration being treated the same as a national application filed on the International Registration date. The non-examination offices have six months to refuse the International Registration, by providing at least a first office action report. The examining offices have a maximum of 12 months, if the contracting party has declared that this limit will apply, to refuse the International Registration.

If the International Registration is deferred, it is kept confidential for the deferment period. This term will depend on the designated contracting parties, and has a limit of 30 months. When the deferment ends, it is published and the national offices start their reviews. In the interim, a copy of the deferred International Registration can be requested by the national offices designated in the International Registration to use in connection with their procedures.

The protection term for each contracting party will be the same as for the national design registrations, subject to a requirement under the 1999 Act of a minimum of 15 years protection from the International Registration date.

These time features show that the International Registration will be processed rapidly in non-examination contracting parties. The examining contracting parties needed more time to complete their work, but there are limits that help to insure efficient operation. The 1999 Act has brought together two different approaches to design protection in a way that allows them to have the benefits of an international filing system and to work together for improved design protection.

CHAPTER 3

Topics – The Geneva Act (1999) History and Analysis

Access to International Registration File

Meeting of Experts and Diplomatic Conference

The experts' discussion of whether there should be access to the file of a published International Registration did not suggest any problems. WIPO proposed that the files copies could be obtained. Before registration, all files would be confidential.

The Meeting of Experts agreement to use specimens initially for two-dimensional designs, instead of reproductions, for a deferred International Registration created a unique situation for most contracting parties. The specimen would be replaced with a reproduction before deferment ended. A concern was expressed at the Third Meeting of Experts that the reproduction might not be the same as the specimen, and some way to compare these items was needed [Fifth Session Final Report #111 (general summary)]. This comment led to a discussion of which International Bureau records would be accessible before the deferred International Registration was published.

Another issue address by the experts was the International Bureau's role in reviewing the reproduction that replaced the specimen. The question was whether the International Bureau would reject the reproduction if it was not for the same design.

These concerns on access to the deferred international file were answered, after considerable discussion of another topic, prior art effect of a deferred International Registration. A copy of the deferred International Registration, including a specimen, would be sent on request to the national offices designated in the International Registration and the inspection would occur there. This solution was accepted by the Diplomatic Conference.

1999 Act Procedures on Access to the International Registration File

It is clear that the International Bureau will keep the International Registration files confidential before publication [1999 Act, art. 10(4)]. When a deferred International Registration is registered, the International Bureau sends a copy of the International Registration to designated offices that have requested copies of these documents [1999 Act, art. 10(5)(a)]. This step provided a way for national offices and third parties to determine whether new matter was included in the reproduction filed after a specimen was replaced by the reproduction. For this reason, each national office that will receive deferred International Registrations should request receipt of an International Registration copy, as this document may need to be in their search files and accessible to third parties.

There are special situations that may allow a third party to have access to a deferred International Registration file [1999 Act, art. 10(5)(b)]. At all times the documents must be kept confidential by the third party.

The International Bureau will provide information on published International Registrations and copies of their files on request, for a fee [1999 Act, rule 32(1)]. There is no definite time length for the International Bureau to retain the International Registration file documents. This policy decision can be addressed by the Hague Union Assembly. The most important questions on access to the International Bureau records were answered in the 1999 Act.

Act of 1934, London, June 2, 1934

General

This treaty was the second of the Hague Agreement treaties [Appendix IV]. Members of the 1934 Act are members of the Hague Union under the Hague Agreement 1999 Act. The 1934 Act had a maximum of 15 years protection, in an initial 5-year plus 10-year intervals [1934 Act, art. 7]. The first protection stage could be a sealed deposit, not opened except under certain circumstances (1934 Act, art. 8]. When unsealed the design deposited was published. The deposited design was forwarded to the designated State offices. There was no apparent rejection of the deposit by the Member States. The 1934 Act included the principle that it would have no effect on certain national design protection rights [1934 Act, art. 21], the 1934 Act merely providing a link to the national systems for filing purposes (1934 Act, art. 4(2)).

Members of the 1934 Act, as of July 1, 2004

The members of the 1934 Act are: Netherlands Antilles, Benin, Switzerland, Côte d'Ivoire, Germany, Egypt, Spain, France, Hungary, Indonesia, Liechtenstein, Monaco, Morocco, Senegal, Suriname, Tunisia, Holy See (17 States). Appendix IX has a chart of the Hague Union Acts and their membership.

Effect of Earlier Treaty Memberships under the 1999 Act

When designating protection in an International Registration application for a State that is a member of the 1934 Act, it is necessary to determine which Hague Union Act controls [1999 Act, art. 31]. When the designated State is a member of the 1934 Act and a subsequent Hague Union Act, the latest subsequent Act controls if both the applicant's contracting party and the designated State are members of that treaty at the time of deposit. For example, Liechtenstein is a member of the 1934 Act, the 1960 Act, and the 1999 Act. An International Registration application selecting Liechtenstein as the originating contracting party will have the 1960 Act applied to Italy, because that is the latest Hague Union Act common at this time. As to Hungary, the 1999 Act will control. Subsequent membership in a later Act will not change the controlling Act for prior registrations. It will be very important to update information on the membership status to determine the applicable Act. The WIPO website will have that information, URL: http://www.wipo.int.

The Hague Act of November 28, 1960

General

The Hague Act of 1960 is part of the Hague Union [Appendix V]. This Act introduced the possibility of a Member State rejecting a deposit, within a six-month maximum time period (1934 Act, art. 8). It was not a sufficient time for most novelty examination countries to process an international deposit, so a major increase in membership of the Hague Union did not occur. The protection term was an initial five years, with five year renewal internals for as long as the national law provided (1960 Act, art. 11).

Members of the 1960 Act as of July 1, 2004

The members of the 1960 Act were: Belize, Benelux, Benin, Bulgaria, Cote D'Ivoire, Croatia, Democratic People's Republic of Korea, France, Gabon, Georgia, Germany, Greece, Hungary, Italy, Kyrgyzstan, Liechtenstein, the former Yugoslav Republic of Macedonia, Republic of Moldova, Monaco, Mongolia, Morocco, Niger, Romania, Senegal, Serbia & Montenegro, Slovenia, Suriname, Switzerland, and Ukraine (29 States). (Appendix IX has a chart of the Hague Agreement Acts and their membership.)

Effect of Later Treaty Memberships

The 1934 Act section on this subject should be consulted.

Act of 1999, Geneva Act of July 2, 1999

General

The 1999 Act is the latest version of the Hague Agreement. It is part of the Hague Union [1999 Act, art. 1 (xxiv)], along with the 1934 and 1960 Acts. There was a need for a global treaty to facilitate industrial design filing. This step also would provide a forum for discussion of design protection improvements. States with novelty examination offices participated in the development of this treaty, increasing the likelihood that all types of national design systems could be accommodated. A number of major compromises were made to allow the States with simple registration systems and the novelty emanation States to achieve benefits. A review of the 1999 Act development is given in the chapter I, General History of the Geneva Act (1999).

The basic structure of the 1999 Act included features from the prior Hague Agreement Acts in updated form. For example, the refusal period for novelty examining systems was adjusted to a realistic limit of 12 months to obtain a report on examination (first office action in the U.S.). A modified approach to deferred publication of the International Registration was adopted, to accommodate States without the possibility of deferment. An arrangement was set up to use deferred International Registrations in the novelty search for examination offices. The prior art effect of an International Registration was clarified. The 1999 Act had worked out a suitable approach to handle the examination of International Registration multiple design applications. These and other improvements achieved a greater possibility of a global industrial design filing treaty.

More details on the above mentioned topics concerning the 1999 Act can be found in the respective sections.

Members of the 1999 Act as of September 3, 2004

The members of the 1999 Act are: Croatia, Egypt, Estonia, Georgia, Hungary, Iceland, Kyrgyzstan, Liechtenstein, Nambia, Republic of Moldova, Romania, Slovenia, Spain, Switzerland and Ukraine (15). (Appendix IX has a chart of the Hague Agreement Acts and their membership.)

Effect of 1999 Act Membership

An applicant designating for protection based on an international application originating from a 1999 Act member can only obtain protection in other 1999 Act contracting parties. If a 1999 Act contracting party is also a member of an earlier version of the Hague Agreement and the 1999 Act, and the designated contracting party is too, then the protection will be based on The 1999 Act. The Act of 1934 section should be consulted for more information on the applicable treaty between contracting parties.

Administrative Instructions – Act of Geneva, July 2, 1999

General

The Administrative Instructions of the 1999 Act provide details on Rules of the Common Regulations. The Administrative Instructions must not be in conflict with the Act or rules [rule 34(4)]. The approval process in the Hague Union is less complex for an Administrative Instruction, compared to a rule change or adoption of a new provision of the Act. Therefore, matters of a generally non-controversial nature and subject to frequent administrative change, or requiring prompt action, are best left for preparation as Administrative Instructions. An Administrative Instruction is prepared by the Director General who consults with all the contracting parties to achieve agreement. These and other aspects of the 1999 Act Administrative Instructions relation to a rule are dealt with in rule 34.

Specific Administrative Instruction Sections

The categories of Administrative Instructions, as of April 1, 2004, included:

1. Communications with the International Bureau, sections 201-204,
2. Requirements Concerning Names and Addresses, sections 301-302,
3. Reproduction of Industrial Designs; Disclaimer; Numbering, sections 401-406,
4. Refusal, sections 501-502,
5. Requests for the Recording of a Limitation or Renunciation when Publication Deferred, section 601,
6. Renewal, section 701,
7. Mode of Payment, section 801.

A regular check for new and modified Administrative Instructions on the WIPO Hague System website would be a wise step (URL: http://www.wipo.int).

Assembly of Hague Union and Administration

Meeting of Experts and Diplomatic Conference

It was not until the Seventh Meeting of Experts that WIPO proposed that the 1999 Act should be in the form of a revision of the Hague Union. Up to that time the experts were considering two approaches, one being a revision of the Hague Agreement Act and retaining the Hague Agreement Union, and the other a treaty separate from the prior Hague Agreement Acts. The existing relationships under the Hague Agreement made it difficult to develop a separate act. There was approval without significant discussion of the WIPO recommendation to provide an updated Act for Hague Agreement Union. This topic was not debated significantly thereafter in the Meeting of Experts or the Diplomatic Conference [Seventh Session, Draft Act, Introduction, #1].

1999 Act Hague Union

Every international agreement needs an administrative structure and, generally, for the WIPO treaties they are similar. The members are formed into a Union, with WIPO, through its International Bureau, providing the administration [1999 Act, art. 22]. Members of the 1934, 1960 and 1999 Acts of the Hague Agreement form the Hague Union. It is helpful to think of the Hague Union in relation to its Acts as a process of updating, with the latest Act the primary one in use. The earlier Acts remain active, while they are used less, and the goal is to have all contracting parties a member of the 1999 Act.

The Union is governed by an Assembly formed from contracting parties that have joined one or more of the Acts that form the Union. In the case of the current Hague Union the Assembly is formed by members of the Complementary Act of 1967 (1999 Act, arts. 20 and 21). The Hague Union has a unique situation for States that are members of only the 1934 Act. These States cannot vote in the Assembly formed under the 1999 Act on matters related to the 1999 Act. They can only be observers. Fortunately, many members of the Hague Union are members of the 1960 and the 1999 Acts, as well as the 1934 Act, and they are voting members of the Assembly on 1999 Act matters.

The 1999 Act set up the voting structure for the Assembly, with voting requirements for adopting and changing the provisions of the Act, regulations and Administrative Instructions (1999 Act, art. 21). These are several critical provisions that have higher voting requirements.

Intergovernmental Organizations

The development of a treaty usually involves some debate on administrative process, and the 1999 Act was no different. The principle issue was the right of intergovernmental organizations to vote separately from their States. This disagreement continued from the Fourth Meeting of Experts to the 1999 Diplomatic Conference [WIPO Records of the Diplomatic Conference ... Hague Agreement (Geneva Act) 512-513 (#1069 – #1080) (2002)]. The solution was to give the intergovernmental organizations the vote to represent all of its States if none of its States voted [1999 Act, art. 21(4)(b)(ii)]. If one of its States voted, the intergovernmental organization could not vote, only its States could vote. This negotiation was with the European Union.

Common Office

Another unique feature of the Hague Union organization was provision for States that are members or plan to be members of the 1999 Act to have a common office handling their industrial design registration (1999 Act, art. 19). These States must share a common legislation system for design registration [WIPO Records of the Diplomatic Conference ... Hague Agreement (Geneva Act) 422 (#261-262) (2002)]. The Benelux Office for the Netherlands, Belgium and Luxembourg is an example of a common office that would qualify under this provision of the 1999 Act.

Special Voting Requirements of the 1999 Act and Rules

The Geneva Act structure is preserved by the General Assembly voting provisions (1999 Act, arts. 25 and 26). On general matters the Assembly must have present a quorum of 50% of the contracting parties to take action by two-thirds of the votes cast [1999 act, art. 21(5)]. If less than 50% of the contracting parties are present, under some circumstances action can be taken with special post meeting validation of the action [1999 Act, art. 21(3)]. Amendment of the 1999 Act can be made by three-quarters of the votes cast, with a few important exceptions [1999 Act, art. 21(2)]. As to 1999 the Act arts. 21 and 26, a four-fifths voting minimum was set for amendment to the Act (1999 Act, arts. 21 and 26). For example, this arrangement makes it more difficult to change the negotiated voting status of intergovernmental organizations.

The 1999 Act Common Regulations can be amended by a vote of two-thirds of the votes cast with a quorum present, with a few important exceptions [1999 Act, 21(2)(iv) and 21(5), rule 33]. Some regulations require a unanimous vote or a four-fifths vote (1999 Act, 24(2)]. The rules where unanimity is required are:

> Rule 13(4) (indirect filing, six months to transmit application for examining contracting party),
>
> Rule 18(1) (refusal period, where examining contracting parties can declare that six additional months, and a total of 12 months applies).

The rules with the requirement for a four-fifths majority of the votes cast with a quorum are:

> Rule 7(7) (multiple designs in an application must be in the same class),
>
> Rule 9(3)(b) (contracting party limit on requesting more than six views for a design in the reproduction),
>
> Rule 16(1)(a) (International Registration deferment maximum 30 months),
>
> Rule 17(1)(iii) (six months before publication of an International Registration).

These voting requirements were the result of negotiations and protected provisions from change that were considered more important by some contracting parties.

Cost to Users

Meeting of Experts and Diplomatic Conference

The Meetings of Experts did not make a recommendation on fees for the International Registration services. There was a fee schedule circulated in several Draft Acts. Cost of services, however, was very much a topic in each experts meeting session, with the emphasis on keeping costs reasonable. In particular, the textile, fashion and other industries stressed the fact they had a large number of designs and the registration process was very expensive under the current Hague Agreement. The following analysis of cost to users is for the 1999 Act. The approved fees are in a schedule attached to the 1999 Act Common Regulations.

One of the 1999 Act accomplishments was to establish a process for handling two-dimensional designs, like clothing fabric design, allowing filing of a specimen of the material to save filing costs. This step, coupled with the deferred International Registration requirement when specimens were use, allowed considerable potential cost saving. The design owner could wait until the important designs became evident to complete the International Registration application process only for these designs. Before deferment ended, a reproduction was substituted for the specimen to complete the application for publication. The fee for filing an application with a specimen was much lower than the regular fee where a reproduction had to be provided. The deferment allowed an applicant to file early and retain secrecy, an important protection in several industries.

Multiple International Registration Fees

The Hague Agreement 1960 Act provided for filing more than one design in an International Registration. The fee structure for the multiple designs allowed a big saving in filing costs over separate applications for each design. The additional fee for each extra design was only a small fraction of the fee for one design application. The rationale was that a lot less work was required by WIPO to process the additional designs, since the application file for each design was the same. The reduced cost for multiple designs applied on renewal also, adding to the cost saving. There were limitations on what designs could be added and how many. Only designs from the same International Classification could be included and there was a limit of 100 total designs in an application (See the section on International Classification for more details on that subject). The 1999 Act used essentially the same cost reduction approach as the 1960 Act for multiple design International Registrations.

Individual Designation Fee

The national office cost for processing varies, from the ones that have novelty examination to a lower cost for the simpler registration type review. For this reason, the 1999 Act allowed a contracting party to have its own designation fee. The individual designation fee cannot exceed the national fee for corresponding work, and it should be less, as WIPO's formal processing comes ahead of the examination office work. The individual designation fees range from CHF 94 to 188 for one design, and the designation fees for additional designs in an application are approximately 25% of the first design cost. A Contracting Party can require the designation fee in two parts or all fees paid at once. This feature allows the U.S. to continue its one fee practice for publication and protection over the full term [WIPO Records of the Diplomatic Conference ... Hague Agreement (Geneva Act) 414-420 (#186 – #206 (2002)].

Normal Types of Fees

The main types of fees are basic and publication. The basic fee is for WIPO initial processing of the International Registration, paid when the application is filed. For an application that will go immediately to publication, the publication fee is paid at the time of filing to save time, since publication will occur in about six months. This normal, fast moving case illustrates the importance of having an account with WIPO to pay fees promptly. The Administrative Instructions, section 801, describe the ways to pay the fees an account with the International Bureau, a Swiss postal check, a banker's check and cash. The basic fee and publication fee include the fees for additional designs in an application.

Deferment as a Cost Saving Step

An applicant may be able to obtain publication deferment for any type of design, depending on which Contracting Parties are designated for protection (*see* the Deferment section for more details on that procedure). Procedurally a deferred International Registration has the effect of a national application in the designated contracting parties on the International Registration date upon acceptance by the International Bureau. The acceptance is based on the International Bureau finding that the application has the required contents. The only step that is not completed is to publish the information on the International Registration. The effect of requesting deferment is to defer some of the fee costs. The publication fee can be paid up to three months before the deferment ends [1999 Act, rule 16(3)]. Since a deferred International Registration can be renounced or limited up to the time of publication fee payment, a business can wait and decide which designs to protect and save the publication costs or withdraw the designs.

Use of Specimen and Deferment as Cost Saving

When deferment is requested, an applicant for protection of a two-dimensional design has the option of filing a design specimen (*see* the Specimen section for more details on that subject). The specimen has to be replaced with a reproduction and the registration fee paid [rule 16(3)]. This step was set up for the textile, fashion and other industries that have many designs and need a low cost way to initiate protection. The use of specimens and deferment saves the cost of initially preparing a reproduction and paying the full publication fee [1999 Act, art 11(5)], when design protection in a contracting party is renounced. A business decision on which designs should be protected can be delayed by up to 30 months in some situations, and significant costs may be saved by effective management of the protection program.

Example of Cost Savings Using Multiple Design International Registration Application

1. Typical International Registration – Multiple Designs

Basic Fee	397 CHF
10 additional designs fee	190
Publication Fee	12
19 additional designs fee	120
Standard Designation fee	42
Total fee	695 CHF
Cost per design	69 CHF (US$ 57)

2. Typical International Registration – Five Year Renewal

One design	216 CHF
10 additional designs	170
Total	367 CHF
Cost per design	35 CHF (US$ 28)

(Exchange rate July 11, 2004 CHF 1 = US$ 0.82)

Contracting Party

Meeting of Experts and Diplomatic Conference. Special Case – Intergovernmental Organization

At the Fourth Meeting of Experts there was a major discussion and difference of opinion over the status of intergovernmental organizations on the right to vote. [Fourth Session, Final Report #145]. This debate continued until the last few hours of the Diplomatic Conference (see Assembly of the Hague Union and Administration for more information on the EU intergovernmental voting arrangement).

1999 Act Use of the Term Contracting Party

The phrase "Contracting Party" is used in several ways in the 1999 Act and plays a key role in the Act's structure. It is used as a general term to embrace both States and intergovernmental organizations that are members of the Act. [1999 Act, art. 1(xiii)]. The intergovernmental group was a necessary addition due to the European Union (EU) plans, initiated during the development of the 1999 Act, for the Community Design Regulation. If the 1999 Act was to have a global effect, it had to link with regional design protection systems.

The term "Contracting Party" was used extensively also in the International Registration application, where the applicant's contracting party had to be identified [1999 Act, art. 1 (xiv); *see* application in Appendix VIII]. The status of the applicant's Contracting party comes from the entitlement that gives a party the right to file under the Act (*see* Topic chapter subject: "Filing – Entitlement", for more information on using the 1999 Act). Each International Registration has only one applicant's contracting party, even though there may be more than one contracting party that could be used. It is selected when the application form is filled out.

In most countries design protection may be possible with several forms of intellectual property rights. A contracting party will designate the office which will implement the 1999 Act [1999 Act, art. 1(xvi)]. For example, in the U.S. this national office will likely be the Patent and Trademark Office and the interfacing protection system will be the design patent system. The only system the U.S. delegations discussed on the record at the Meetings of Experts and the Diplomatic Conference was the design patent to implement the 1999 Act. Later, other U.S. design protection systems could be added by appropriate legislation.

International Registration Application Form – Contracting Party

A review of the International Registration application form [DM/1(E) – Appendix VIII] is a helpful way to illustrate the various uses of the contracting party concepts. Under Item 3, Entitlement to File, each applicant must state the following:

1. where the applicant is a national,

2. a contracting party where the applicant is domiciled; and
3. a contracting party where the applicant has a real and effective industrial or commercial establishment.

Each of these items of information must be stated, if there is a basis for the status. For 1999 Act members, an applicant must state a contracting party where the applicant has a habitual residence.

In Item 4 of the application, each applicant must indicate the "Applicant's Contracting Party". This designation will control matters that are affected by the contracting party relation for that application, such as which contracting parties can be designated.

In Item 10 the "Designated Contracting Parties" for which protection is sought will be selected.

Contracting Party Obligation to Report Invalidations

A contracting party has a number of administrative details to report regularly to the International Bureau. There was discussion at the Diplomatic Conference on the contracting parties' duty to report International Registration invalidations. It was concluded that only invalidations known to the contracting party office need to be reported [WIPO Records of the Diplomatic Conference ... Hague Agreement (Geneva Act) 455 (#531 – #534].

Contracting Party Implementation of the Geneva Act (1999)

At the Diplomatic Conference the Irish delegation pointed out that its Constitution only permitted provisions in Irish statutes to be effective in Irish law. The Irish delegation was concerned that the Draft Act would not be implemented in Ireland, as it would bind the country to a law not in their statutes. An amendment to art. 2 of the Draft Act was made to accommodate the Irish situation [WIPO Records of the Diplomatic Conference ... Hague Agreement (Geneva) 507-508 (#1036 – #1039) (2002)]. This change raises interesting issues of international law concerning treaty effect and implementation that should be studied by each contracting party.

Declarations

General

The concept in developing the 1999 Act was to minimize the need for members to change their design registration systems. In addition, the 1999 Act was not intended to be the cause of substantive law harmonization. The goal was to achieve an effective international filing system, attractive to all countries and intergovernmental organizations. The 1999 Act provided for exceptions that could be made by a member, through declarations or notices (generally referred to here as declarations). These changes allowed the 1999 Act to interface with the national design registration system. In most cases the exception was allowed only for contracting parties with examining offices.

Every country and intergovernmental organization considering membership in the Hague Agreement under the 1999 Act needs to review these exceptions, to determine what declarations to make. Most of these actions can be taken at the time of joining, and at other times when the need arises [1999 Act, art. 30].

This section is a check list and explanation of the declarations that may be made under the 1999 Act. The declarations are presented here in alphabetical order. More background on some of the declarations can be found under the related section. All these declarations must be submitted to the WIPO Director General, and the notices are sent to the International Bureau.

Check List of Declarations and Notices

ADDITIONAL MANDATORY CONTENT OF INTERNATIONAL REGISTRATION – All International Registrations must have mandated content to receive a filing date (1999 Act, art. 5(1)]. A declaration must be made by examining offices to require additional content in an International Registration for a filing date in that contracting party [1999 Act, art. 5(2), rule 18(1)]. A list of permitted, additional content is found in 1999 Act, art. 5(2)(b). The additional content items that can be declared are: identification of the creator, a brief description of the industrial design, and a claim. In the declaration, if a claim is required, the exact working of the claim must be given. If the creator's name is required, the declaration must state the acceptable form and content for the creator information [1999 Act, rule 11(3)].

COMMON OFFICE – If two or more States have or are planning to unify their industrial design registration system and form a common office, a declaration must be filed to fall under the special provisions of the 1999 Act [1999 Act, rule 6(3)(ii)].

COMMUNICATIONS BETWEEN INTERNATIONAL BUREAU AND NATIONAL OFFICE – A national office needs to notify the International Bureau if it wants all communications with WIPO in one language, either French or English [1999 Act, rule 6(3)(ii)].

CONTRACTING PARTY DENIAL – USE OF OFFICE FOR INDIRECTLY FILING INTERNATIONAL REGISTRATION – A Contracting Party must submit a declaration if it does not want International Registrations to be filed indirectly through its office [1999 Act, 4(1)(b)]. With the advent of a successful Patent

Cooperation Treaty that uses indirect filing, it would seem that this option should be allowed and properly managed. For the U.S. it would be a major improvement, providing a streamlined way to comply with the national security review and obtain foreign design protection.

DEFERMENTS – At the time a contracting party deposits accession to the 1999 Act, a declaration must be made on the time period allowed for deferment [1999 Act, arts. 11(1)(a)]. If the national laws do not permit any deferment, this information must be provided [1999 Act, arts. 11(1)(b)]. A contracting party may have no deferment period. The no deferment declaration was needed, since at least the U.S. indicated that it could not agree to deferment of a design publication.

DESIGNATING APPLICANT'S CONTRACTING PARTY – A contracting party may submit a declaration that no protection can be designated in that contracting party by applicants basing their filing entitlement in that contracting party, also referred to as the "Applicant's Contracting Party" [1999 Act, art. 14(3)].

EFFECT OF INTERNATIONAL BUREAU OWNERSHIP TRANSFER RECORDING. A contracting party may provide in a declaration that an International Registration transfer of ownership recorded at the International Bureau would not have effect until the office of the contracting party has received the specified document [1999 Act, art. 16(2)]. This step was specifically requested by the U.S., to help preserve the centralized integrity of its ownership records related to International Registrations.

INDIVIDUAL DESIGNATION FEES – A contracting party with an examining office may file a declaration setting the individual designation fee for examination, publication, renewal, and for payment separately or combined to include renewal fees for the full protection term [1999 Act, art. 7(2)]. This Declaration was needed to allow the U.S. to continue its practice of requiring one payment for publication and the full protection period.

NATIONAL OFFICE COLLECTING AND TRANSMITTING FEES TO WIPO [1999 Act, rule 27(2)]. If a national office plans to collect fees concerning applications filed indirectly through their office and transmit the fees to WIPO, the Director General must be notified.

NATIONAL OFFICE TRANSMITTAL FEE – If a national office will charge a transmittal fee to send the indirectly filed International Registration application to the International Bureau, that office must notify the International Bureau what fee will be charged [1999 Act, rule 13(2)]. The fee must be only for the costs incurred.

PROTECTION DURATION – Each contracting party must submit a declaration stating the maximum term of protection for the national design registration that will be the basis for the International Registration protection [1999 Act, art. 17(3)(c)]. This information will be used to set the maximum protection term for the International Registration.

REFUSAL PERIOD EXTENSION. A contracting party with an examining office may file a declaration to extend the period of refusal by six months, for a total of 12 months [1999 Act, rule 18(1)(b)]. This declaration was the result of extensive negotiations, with several examining countries to reduce their examination time, and to clarify what must be provided in the refusal statement.

REQUEST FOR COPY OF DEFERRED INTERNATIONAL REGISTRATION A confidential copy of a deferred International Registration will be sent to each contracting party designated in a deferred International Registration, if a general notice is given to provide these copied [1999 Act, art. 10(5)]. This document is essential for examination offices to evaluate prior art, since the International Registration may be a prior art document from the date of the International Registration, or earlier.

REPRODUCTION – VIEWS REQUIRED – A contracting party must make a declaration if specific reproduction views are required and indicated how these requirements can be met [1999 Act, rule 9(3)]. The International Registration application form instructions will incorporate this information.

SECURITY REVIEW – Where a contracting party requires a national security review of design applications that are foreign filed, a declaration may be made to have the period for transmitting an indirectly filed International Registration to the International Bureau changed to six months [1999 Act, rule 13(4)]. The special situation in the U.S. for national security review required that this provision be included, so that the U.S. would have additional time to mail the indirectly filed International Registration to the Bureau.

TIME OF EFFECT FOR INTERNATIONAL REGISTRATION, A contracting party that has a examining office may declare when the legal effect of an International Registration will occur [1999 Act, art. 14(2), rule 18(1)(c)]. It was quite clear from the experts meetings and the Diplomatic Conference that the U.S. required some flexibility to control this date, so that its records could be properly kept and carefully controlled when the rights under an International Registration commenced.

UNITY OF INVENTION – A declaration must be filed if a contracting party's law has the requirement for unity of invention limiting protection in an International Registration to one design concept [1999 Act, art. 13(1)]. This flexibility was needed as some countries allow only one design concept in a registration. The solution was to treat the International Registration with several design concepts in the same manner as national applications, and allow division to set up the necessary individual applications with the original filing date.

Deferred Publication

Meeting of Experts

The history of industrial designs protection started with the concern over access to the deposited designs (French system of sealed design deposit in Lyon, France and the UK industrial design registration system in 1841). Initial secrecy was essential in some industries. Many countries follow this tradition now. The 1934 and 1960 Acts of the Hague Agreement had provision for deferring publication of the International Registration [1934 Act, arts. 8 (5 years deferment); 1960 Act, art. 6(4) (12 months deferment)].

One of the most discussed topics at the early Meetings of Experts on the Hague Agreement revision was deferment [*see* chapter I, General History of The Geneva Act (1999) for more background]. A proposal by the Swiss delegation settled the matter [Third Session, Draft Act art. 7, notes 7.01-7.09 to art. 7, Final Report # 103-116]. The solution was to follow the mandate of the Hague Agreement that national laws would not be changed, unless necessary. If a country designated in an International Registration had a different deferment period than the 30-month maximum provided in the Draft Act, the shortest deferment period would control. For example, if Switzerland had a 30-month deferment period and Iceland had no deferment period, then an International Registration designating these countries would not have any publication deferment. It would be published by the International Bureau upon approval six months from the International Registration date.

According to this plan the International Registration rights would be obtained upon registration, whether the International Registration was published or deferred. It followed the classic approach of prior systems, going back to the French sealed deposit and the 1934 and 1960 Acts. When the deferment period was over the International Bureau would publish the reproduction and other information on the registration.

The experts recognized that deferment had several advantages, including the confidentiality of the design for a limited period. Initial protection costs could be reduced using deferment. This cost reduction advantage became more evident as other features of the Hague Agreement revision were developed.

1999 Act Deferment Procedure

The Diplomatic Conference adopted the deferment procedure set up by the experts, a part of which was that a limited number of contracting parties would receive copies of deferred International Registrations [1999 Act, art. 5(5) and 11, and rule 10]. The Diplomatic Conference decided that any of the designated contracting parties could receive a copy of the deferred International Registration, if a general notice to receive copies was given. The importance of these documents was recognized.

The Diplomatic Conference supported the approach to continue without deferment an International Registration application, even though the application requested deferment and designated a contracting party that did not allow deferment. User groups expressed the view that retaining the protection was the essential step [WIPO Records of the Diplomatic Conference ... Hague Agreement (Geneva) 419-420 (#232 – #241 (2002)]. Several features of the 1999 Act interact with deferment, and the reader is encouraged to review the related sections for additional background.

The 1999 Act International Registration initial processing by the International Bureau was the same for all applications. As shown in the current application form [Appendix VIII], there is a place in sec. 13 to request deferment, in accordance with 1999 Act, art. 5(5). The number of months requested for deferment must be stated. The form notes for sec. 13 provide up-to-date information for limits controlling this selection. The designation of contracting parties for protection control how many months, if any, will available, and the contracting party with the shortest deferment period will be the deferment time allowed for that registration.

The fee for publication is not paid on filing when a deferment is requested (1999 Act, rules 12(2), 16(3)(a), introducing another cost saving factor. The publication fee must be paid no later than three months before deferment ends.

Deferment and Specimen Filing

Where deferment is requested for a two-dimensional design, an applicant can substitute a specimen of the design and received a filing date. [1999 Act, art. 5(1)(iii) and rule 10]. For example, a textile company can submit 100 different designs in an International Registration, since they are in the same International Class, filing specimens of each fabric design. In this multiple design International Registration the specimens would have to be replaced by reproductions before publication (1999 Act, rule 16(3)].

The effective date of the deferred International Registration is the International Registration date, and the International Registration will be treated as a national application filed on that date [1999 Act, arts. 9, 10(2) and 14]. The fact that there was deferment and no publication of the reproduction until the deferment period ends does not alter the filed national application effect given the International Registration.

Six Months Mandatory Deferment – All International Registrations

All International Registrations are given a six-month deferment from publication from the International Registration date, unless earlier publication is requested by the applicant [1999 Act, rule 17(1)]. This time period corresponds to the minimum refusal period of six months, after which the International Registration protection commences, from the International Registration date for non-examining contracting parties [1999 Act, art. 14(2) and rule 18(1)(a)].

Confidentiality of Deferred International Registration

The deferred International Registration is held in confidence by the International Bureau, with provision for access under certain circumstances [1999 Act, art. 10(4)]. The International Registration disclosure during deferment was made very limited. The disclosure must be in confidence under all circumstances [1999 Act, art. 10(5)(b)]. This feature was strongly emphasized by the contracting parties relying on deferment. There should not be any way for competitors to gain access through a non-confidential disclosure for litigation or other proceedings.

National Office Receipt of Deferred International Registration

The Diplomatic Conference discussed which contracting parties should receive the confidential, deferred International Registration. The experts' proposal in the Draft Act was to have only designated examining contracting parties receive the document. The Diplomatic Conference decided all designated offices would receive a copy, if notification is given to the International Bureau that they want to receive copies [1999 Act, art. 10(5)(a), WIPO Records of the Diplomatic Conference ... Hague Agreement (Geneva) 439-442 (#401 – #428 (2002)]. The discussion had only a few general references on how a contracting party examining office would use the deferred International Registration document in the examination process, or what the prior art effect might be for a deferred International Registration (*see* section on International Registration – Effect).

At least the designated examining offices need to receive a copy to use in their examination, if there is prior art effect due to the deferred International Registration. The national offices are obligated to keep the copies confidential, even in connection with their proceedings.

National Effect of a Deferred International Registration

The fact that a deferred International Registration is not published creates unique issues as to rights and prior art effect that each contracting party must address under their laws. Many countries have dealt with the deferred publication situation for a long time, where rights existed and there was no knowledge of what the design was for a relatively long period.

An International Registration effect after registration, in general terms, is answered by looking to the national laws. The question is how would a deferred national application filed in that country be treated? The answer for the International Registration is the same as the effect of a deferred national application, unless there are special circumstances dictated by the treaty. In some countries the International Registration, upon approval, may be treated as a registration at the time of filing, on the International Registration effective date, with rights to protection created at that time. Other countries may reject the concept of deferral and they may have an examining system that will determine whether rights should be granted, only after the examination is completed and registration is approved. These countries may have an effective deferred publication of the design until that examination is completed and there may be rights that go back to the filing date after the examination is completed. The section on International Registration Effect should be consulted for more details on this subject.

Description of the International Registration Design

Meeting of Experts and Diplomatic Conference

One major difference in industrial design practice around the world is the requirement for a description of the design in some systems. Other systems rely primarily on the picture of the design to determine the protection scope. At the Meetings of Experts there were extensive discussions on whether to require, or even allow, a design description [*see*, for example, the Fifth Session Final Report, items 44-48]. The general view was that the new treaty should not have a mandatory requirement for a design description and that the design picture should be the primary source for determining the protection scope. There was considerable concern expressed that any description of a design, title or requirement for an indication of a design product or use, might inappropriately narrow the protection scope. Others expressed the view that at times a description could be used to help focus on features that identify the novel design aspects.

The Meeting of Experts and the diplomatic conference discussions made it clear that a description had to be of limited content and size. It was not a substitute for the visual impression of the design picture.

1999 Act Approach to International Registration Design Description

The approach taken in the 1999 Act on whether to require a design description followed the general concept of the Hague Agreement – a contracting party joining the treaty should not have to change its procedures, unless necessary for the treaty to be effective. A contracting party can declare that a description is required to obtain a filing date for an International Registration designating that contracting party for protection [1999 Act, art. 5(2)(b)(ii), rule 7(4) and 11(2)]. Otherwise, it is optional whether a description is in the application.

The description cannot include technical features or uses of the design. It must relate to the design's visual features [1999 Act, rule 11(2)]. Also, it is limited to 100 words, and to make that point firm, the fees are CHF 2 per word over 100.

Filing – Entitlement to File an International Registration

Meeting of Experts

Since a decision was made at the Meetings of Experts to have a Hague Agreement revision and not a separate treaty, it was logical that the same principles of when a party was entitled to file under prior Hague Agreement Acts would be followed. These principles incorporated international practice. The 1960 Act principles on entitlement were adopted in the 1999 Act with one addition principle on habitual residence as the basis for filing.

1999 Act Filing Entitlement Principles

The question is – who can file an International Registration under the 1999 Act? The basic principles used are the same, essentially, for many WIPO treaties. They look at the party filing, either an individual or an entity and ask the following questions:

1. Is that party a national of a contracting party of the 1999 Act;
2. Is that party domiciled in a contracting party of the 1999 Act;
3. Does that party have a habitual resident in a contracting party; and
4. Does that party have an effective industrial or commercial establishment in territory of a contracting party of the 1999 Act [1999 Act, art. 3]?

If the answer to any one of these questions is yes, then the party can file an International Registration under the 1999 Act.

The interpretation of these principles on the facts of each situation will be decided under national law of a contracting party. The nationality, domicile and business criteria have been applied under the 1960 Act. The habitual resident criteria is new to the Hague Agreement Acts. It may differ only slightly from the domicile principle, perhaps a little broader in scope.

The effective industrial or commercial establishment criteria has been used by several companies to file under the Hague Agreement. For example, Sony in Japan and Goodyear in the U.S. have filed under the 1960 Act, based on their business establishments in member countries. The type of establishment that would qualify should be substantial. Consideration of related business matters, like tax laws, ownership and licensing, for each country involved would be a wise step.

International Registration Application Form Completion on Entitlement

On the current International Registration application form, Item 3 is where the entitlement to file information is entered [*see* Appendix VIII for copy of current form]. The party filing must identify a contracting party, if appropriate, where the party is a national, domicile and has an established business, and each of these entries are required to be completed with "none" entered if there is no contracting party. A 1999 Act member must indicate a contracting party where there is a habitual residence, or state "none". Separately in Item 4, the party indicates a contracting party bound by the 1999 Act where the party is entitled to use the 1999 Act. This contracting party, if available, is used as the legal basis for the application, and referred to as the "Applicant's Contracting Party". The earlier contracting party entries were a preliminary step to let the applicant identify contacting parties that could be relied on, leading up to the mandatory choice of the 1999 Act, if available. This procedure will result, over time, in the 1999 Act becoming the common treaty between contracting parties.

Filing – International Registration Application

Meeting of Experts

Probably more time over the seven Meetings of Experts was spent on International Registration application filing details than any other topic. By the time the Diplomatic Conference started most of these filing procedures were settled. An understanding of the filing framework is very important.

The primary concern at the Meetings of Experts was to retain the efficient, current Hague Agreement approach for non-examination offices, while allowing examination offices to retain their essential features with commitments to efficient operation.

The tack used at the Meetings of Experts was to reduce as much as possible the mandatory contents for obtaining a filing date. Concessions were made, and in the end only a few exceptions were needed. The first step in harmonizing these procedures was to identify the differences. Next it was necessary to allow for flexibility, and finally an effort was made to negotiate for uniformity, as much as possible.

1999 Act Framework for International Registration Application Filing

The mandatory content requirements to obtain a filing date for contracting parties with non-examination and examination offices are:

1. Request for International Registration under the 1999 Act [1999 Act, art. 5(1)(i)];
2. Information on applicant [1999 Act, art. 5(1)(ii)];
3. Design reproduction (one or more) or specimen under certain circumstances [1999 Act, art. 5(1)(iii), rule 9];
4. Indication (type) of product involved or relationship to a product [1999 Act, art. 5(1)(iv), rule 7(3)(iv)];
5. Designated contracting parties where protection was sought [1999 Act, art. 5(1)(v);
6. Fee [1999 Act, art. 5(1)(vi)]; and
7. Any other prescribed particulars [1999 Act, art. 5(1)(vii)] (*see* below re examining contracting parties).

Contracting parties with examining offices have the option, upon declaration, to add one or more of three additional content items, if they are required under national law, for a party to obtain a filing date in that contracting party. These items are:

1. Identifying the design creator [1999 Act, art. 5(2)(b)(i)];
2. Brief description of the design [1999 Act, art. 5(2)(b)(ii)]; and
3. A claim [1999 Act, art. 5(2)(b)(iii)].

In the case of the U.S., for example, the creator identification and a claim would be required in the International Registration, in addition to the other common mandatory content items. A rule change under U.S. law to allow a filing date without a claim included would be practical only if a general claim such as "what is claimed is the design as shown" was used. The Meeting of Experts negotiations had created a short common list, one that could not be expanded, with a list of the exceptions for examination offices.

Application Form for International Registration Filing

The mandatory International Registration application content requirements to obtain a filing date are clearly identified in the application form [Appendix VIII]. In Item 6 of the form the identity of the creator can be included. As of July 7, 2004, only Bulgaria and Iceland required that information. Other contracting parties could include that information even it was not required for the contracting parties designated.

Item 9 in the form is the place to enter the design description, if it is required by an examination office. As of July 7, 2004, no contracting party has required that a description be included.

There is no place in the current form for a statement of the claim, which is one of the additional items that may be required by an examining office. At the present time none of the contracting parties requires a claim. The claim content requirement will be added to the form at the appropriate time. Included with that request will be the correct claim statement to use, information that will be supplied by the contracting party upon declaring a claim necessary.

The International Registration application form has been prepared to guide the applicant in preparing it. The current information on what is required by each contracting party is shown on the form. Up-to-date application forms, needed as new members join the treaty, should be obtained from the WIPO website, URL: http://www.wipo.int.

Filing – Irregularities in International Registration Application

Meeting of Experts and Diplomatic Conference

There was early general agreement that the procedures in place under the 1960 Act would be used for the 1999 Act concerning how to handle irregularities in application filing.

1999 Act Procedures on Handling Application Irregularities

The International Bureau promptly reports any irregularities in an International Registration application and invites completion [1999 Act, art. 8(1)]. If the irregularity is for a mandatory item to obtain a filing date, then the filing date will be the date the application is completed. If the omission does not effect the filing date, the applicant is given a time period to complete the required information. If the information is not supplied in the time period required, the application will be considered abandoned [1999 Act, art. 8(2)(a)].

When an irregularity effects compliance with a designated contracting party's requirement for a filing date, the International Bureau will delete the designation for that contracting party and proceed with the registration for the other contracting parties [1999 Act, art. 8(2)(b)].

These procedures were set up with the goal of expediting the processing of the application. There is every indication that the prompt correction of irregularities will be facilitated by the International Bureau staff.

Harmonization of Procedural and Substantive Industrial Design Laws

Meeting of Experts

It was clear from the start of the Meetings of Experts that the Hague Agreement revision was not the place to initiate major revision of national procedural and substantive industrial design laws, even though that effort would be a worthy goal. The Hague Agreement revision was to set up a improved central filing system and unify filing procedures, primarily. It was to be compatible as much as possible with existing national design registration procedure and substantive laws.

A few major issues related to procedural and substantive law harmonization were raised at the Meetings of Experts and not included in the 1999 Act. These major issues were grace period, unregistered design protection and design reproduction standards. Comments on a few of the major topics introduced will be presented here, to aid future efforts in harmonizing procedural and substantive design laws.

The 1999 Act did include provisions on protection term and International Registration effect. In the 1960 Act protection term was included, and International Registration effect had been a part of each Hague Agreement Act. There was significant effort to start harmonization of reproduction procedures, and some accomplishing was made, but most of that work remained to be completed. Chapter I, General History of The Geneva Act (1999), and the Topics chapter under the heading "Reproductions" provide additional insights. Another 1999 Act substantive law provision was on International Registration effect, including prior art effect that is central to national design law. The Topic chapter under the heading "International Registration Effect" has a detailed discussion of this subject.

Grace Period

The Second Meeting of Experts discussed adding a 12-month grace period to the Draft Act [Draft Treaty Introduction, #14]. Several countries have grace periods for patents and design registrations, and the Paris Convention has a grace period for exhibitions that is included in the 1999 Act [1999 Act, rule 7(5)(d)]. The principle is that the right to have a patent or design registration is not lost if the owner has published, or taken other commercial steps, when the owner files the application within designated period from the date of the owner's publication or commercial step.

The Second Meeting of Experts concluded that a grace period would not be included in the Draft Act [Second Meeting Final Report, #4]. It was considered a substantive law issue. The Diplomatic Conference followed this recommendation. It is reasonable to protect a design owner against loss of protection at the time business activities may prevent attention to filing an application for registration.

The grace period is one way to prevent that injustice. Another option, unregistered protection, with better results is discussed below.

Unregistered Protection

The addition of unregistered initial protection to the Draft Act was recommended for discussion at the Meetings of Experts, but it was not considered [Second Session Final Report #15]. The Diplomatic Conference followed this recommendation. The unregistered design approach would provide protection from the time a product with the design was made public, without any need for registration. The fact that the European Union Community Design Green Regulation had an unregistered design feature under discussion at the time of the Hague Agreement revision meetings probably motivated the suggestion. The Community Design Regulation had a short period of unregistered protection against copying. Since the EU Community Design Regulation is now law, reconsideration of an unregistered design protection complementary feature for the 1999 Act registration system would be a useful step.

An unregistered protection feature prevents loss of the design right and adds more. The law would eliminate the need to rush to the design registration office, even though the design has been made public. The U.S. has a similar approach in its Vessel Hull Design Protection Act, 17 U.S.C. sections 1301-1332 (2004). Experience with the Community Design Regulation will be a good teacher on whether unregistered protection has benefits.

Indirect Filing (through National Office)

Meeting of Experts and Diplomatic Conference

WIPO did not put indirect filing of an International Registration in the first Draft Act for the Second Meeting of Experts. The procedure would allow filing of an International Registration in a contracting party national office and receiving a filing date, if the International Bureau received the application within a set time. The reason stated for not including it was that little use occurred of that procedure in the 1960 Act [Second Session Draft Treaty Introduction #7]. The experts suggested indirect filing through a national office should be added. At the Third Session a majority favored indirect filing. WIPO wanted minimum delay in receiving the application from the national office.

Another reason for supporting indirect filing was expressed by the U.S. delegation at the Third Meeting of Experts [Third Session Final Report #17]. The U.S. needed to provide a national security review of all design patent applications. Under U.S. law, before a design patent application could be sent to a foreign office a license was needed after national security review. Usually these licenses could be received in a few months, but delays could occur. Since Paris Convention filing for design patents was limited to six months from the first filed national application on that design, the timing presented a tight schedule. The national security review process could not be changed to simplify the design patent interface with the Hague Agreement. Therefore, indirect filing was a good step, as it allowed the U.S. Patent and Trademark Office to conduct the national security review as part of the indirectly filed application processing. [Fifth Session Final Report #34 and #35].

In the Sixth Session, the Draft Act included a proposal of three months where a national review procedure was used to transmit the International Registration to WIPO, plus three more months in more special cases requiring individual notice to the International Bureau [Sixth Session, Draft Act notes 4.02-4.10]. Finally the Diplomatic Conference resolved the issue at the urging of the U.S., to a maximum period of six months, if requested by declaration [WIPO Records of the Diplomatic Conference ... Hague Agreement (Geneva) 442-443 (#429-438) (2002)]. It is likely that the U.S. will ask for this additional time exception to make the interface with the 1999 Act workable [WIPO Records of the Diplomatic Conference ... Hague Agreement (Geneva) 406 (#118-#120) (2002)].

In their Diplomatic Conference opening declarations and later at the conference, Japan and the Republic of Korea expressed a desire not to allow self designation, also stated as designating applicant's contracting party. They explained it would be a major burden on their offices to receive so many International Registration applications in French or English. Later at the Diplomatic Conference these countries requested that a contracting party have the option to prevent self designation, and this provision was included [1999 Act, art. 14(3), WIPO Records of the Diplomatic Conference ... Hague Agreement (Geneva)391-393, 370-371 (#19, #24, #370-378 (2002)].

1999 Act Indirect Filing Procedures

An International Registration can be filed directly with the International Bureau, or indirectly through a national office [1999 Act, art. 4(1)]. A contracting party can make a declaration that no indirect filing can occur in their office [1999 Act, art. 4(1)(b)]. The only parties that can file indirectly in a national office are ones meeting the entitlement requirements of art. 3, a national, or a person domicile there, or having a habitual residence there, or an effective business office [1999 Act, art 4(1)]. A party filing in a national office can designate that contracting party for protection unless the contracting party makes a declaration prohibiting filing [1999 Act, art. 14(3)].

The national office receipt date of a complete International Registration will be its International Registration filing date, if the application is received by the International Bureau within the prescribed time [1999 Act, rule 13]. The International Bureau must received an International Registration application within one month of its receipt at the national office for the national office filing date to be used, unless there is a declaration for an exception due to national security review [1999 Act, rule 13(3)].

National offices with national security clearance reviews can have up to six months to transmit the application to the International Bureau, if a declaration requesting that exception is made [1999 Act, rule 13(4)]. Since this mailing process has very tight time limits, there was considerable discussion on the effect of postal delays. The 1999 Act addressed this concern (1999 Act, rule 5] (*see* the section on Postal Delays for more details).

Interfaces with IP National Laws and Other Treaties

Meeting of Experts and Diplomatic Conference

How the revised Hague Agreement would effect other intellectual property (IP) national laws and treaties was discussed several times at the Meetings of Experts. The concept of how the Hague Agreement interfaced with national law was very important to this discussion. The treaty sets up a separate, international protection system. The procedures and rights are international law and controlled by the treaty. Where the interface occurs is that the treaty provides that the national laws and procedures apply within the bounds of the treaty.

In relation to other IP treaties, it was the intent of the experts that the revised Hague Agreement have no effect on other IP rights obtained in a contacting party. A provision in the 1960 Act had the same result [1960 Act, art. 18]. Therefore, a separate provision was included in the revised Hague Agreement to assure that there would be no conflict with certain existing treaties.

The Fourth Meeting of Experts discussed the treaty interface issue [Fourth Session Final Report #s 140 and 141]. WIPO offered to consider how to clarify the interface, and in the fifth Meeting WIPO staff pointed out that the definition of "Office", in Draft Act, art. 1, helped to clarify the limited impact of the Draft Act on national laws and related treaties. It was only the laws administered by that office that were interfaced with the revised Hague Agreement [Fifth Session Final Report #112]. The U.S. delegation identified the Patent and Trademark Office (PTO) as the U.S. agency responsible for design patents that would be the basis for the interface with the Draft Act. It was clear that other IP forms, such as trademarks, registered by the PTO, and copyright law administered by the U.S. Copyright Office would not be interfaced or affected by the Draft Act. The Seventh Meeting of Experts made this point even clearer. A provision specifically stated that no rights obtained through TRIPS would be effected. This approach satisfied the experts, and the Diplomatic Conference adopted it [Seventh Session Draft Act, art. 2].

At the Diplomatic Conference the UK delegation had a question on the "supremacy" of an International Registration with respect to a national right. The discussion that followed provided addition insight on the interface between the Geneva Act and national law. The WIPO staff explained the concept of the Draft Act, art. 2(1), as to "supremacy" with the following response: "Mr. CURCHOD (WIPO) stated that a national and international registration had to be put on equal footing. The aim of this provision was to make clear that protection for industrial designs granted by national legislation, whatever the basis for such a protection might have been, should not diminish or interfere with the enjoyment of the rights afforded by the new Act. This provision was aimed at preventing national laws from providing less than was conferred by the International Registration system (for example, by providing for a shorter term of protection)" [WIPO Records of the Diplomatic Conference ... Hague Agreement (Geneva) 404 (#101) (2002)]. The Diplomatic Conference discussion was consistent with the WIPO interpretation on this point.

1999 Act Interface with National Laws and Other Treaties

The basic understanding of the Hague Agreement interface with national laws and other treaties discussed by the experts was followed by incorporating a broad interface statement [1999 Act, art. 2]. At the same time the 1999 Act did not preclude contracting parties from designating its other design protection systems from being interfaced with the 1999 Act, even though the 1999 Act did not require the relationship. For example, in the U.S., where separate protection of designs under the Vessel Hull Design Protection Act is provided, with unregistered and registered rights [17 U.S.C. secs. 1301-1332 (2004)], an interface could be made. As of this time, no intention by the U.S. to make this linkage had been indicated.

International Bureau

General

The International Bureau referred to in the 1999 Act is the World Intellectual Property Organization (WIPO) staff in Geneva, Switzerland, that administers the treaties in the Hague Union. It includes a very experienced staff. A very good relationship has developed between users and Bureau staff. If questions arise on an application, the staff contacts the user in an appropriate way to resolve the problems quickly.

As of July 7, 2003, the Trademarks, Industrial Designs and Geographical Indications Group, under Assistant Director General Ernesto Rubio handle all aspects of the 1999 Act operation.

There is no formal appeal department at WIPO to settle disputes on International Registration processing. WIPO uses its internal management structure to review disputes [WIPO Records of the Diplomatic Conference ... Hague Agreement (Geneva) 416-417 (#209-214) (2002)].

International Classification (Locarno Agreement)

1999 Act International Design Classification

National Offices have industrial design classification systems to organize their design registrations. Most of the National Offices use the International Classification, based on the Locarno Agreement, administered by WIPO. The 1999 Act International Registration system uses the International Classification [1999 Act, rule 1(viii)] for several important purposes. Part 8 of the International Registration application (Appendix VIII), has a place to include the International Classification for the product associated with the registered design. If the International Classification is omitted, the International Bureau will add it, and there will be not loss of filing date for that omission.

Access to a copy of the International Classification can be obtained on the WIPO website, URL: http://www.wipo.int, as part of the Hague Agreement System documentation, and national registration offices provide access to the International Classification. Also a very useful WIPO publication, in English, entitled: *International Classification for Industrial Designs (Locarno Classification)*, Eighth Edition WIPO Publication No. 501(E)(2003) is available from WIPO. This publication has the aforementioned product lists and the Locarno Agreement text, along with other useful information. If further help is need in finding the International Classification of a product, WIPO has Locarno Agreement classification experts that may be consulted. A list of classes in the Locarno Classification is in Appendix XI.

Identification of the proper international class should not be difficult for most products with the use of the abovementioned indexes. The International Classification uses the concept of classifying primarily by the purpose of the product associated with the design. A secondary approach is to classify by product shape.

The International Classification is under regular review for improvement. A Committee of Experts meets to evaluate problems and propose changes. Membership in the Locarno Agreement, the treaty associated with this classification system and participation in the Meetings of Experts is a good idea for 1999 Act members.

A 1999 Act member need not use the International Classification in their national industrial design system. Several countries, like Japan and the U.S., have their own classification systems. Even if a member uses a different classification system, it is useful to incorporate the International Classification along with the national classification in each industrial design registration document. National use of the International Classification prepares the application for foreign filing in most national offices and under the Hague Agreement. For example, U.S. design patents have had both the U.S. classification system and the International Classification since December 1996.

Multiple Design Applications

An important role played by the International Classification of designs in an International Registration is as a limitation on the designs that can be included in the application. A maximum of 100 designs can be in the application, and all these designs must be in the same International Classification [1999 Act, rule 7(3)(v) and rule 7(7)].

Legal Effect of the International Registration International Classification

It is up to the national industrial design laws to determine the legal effect of the associated classification system. The Locarno Agreement makes this point quite clear in its article 2(1),which states: "Subject to the requirements prescribed by this Agreement, the International Classification shall be solely of an administrative character. Nevertheless, each country may attribute to it the legal scope which it considers appropriate. In particular, the International Classification shall not bind the countries of the Special Union as regards the nature and scope of the protection afforded to the design in those countries" [International Classification for Industrial Designs (Locarno Classification) viii (Eight Edition WIPO Publication No. 501(E)(2003)].

The practical effect of the International Classification is well known to national industrial design practitioners. In some countries protection is limited to the products in the class or classes designated. Other countries, like the U.S., do not restrict the protection to the class identified in the design registration. In the U.S. if the registered design is incorporated into other products than the one shown in the registration, there still may be infringement, assuming the requirements for infringement are met. This topic is one that needs to be addressed in the international harmonization of industrial design practice. For use of the 1999 Act, the consideration of whether to file more than one International Registration application will be one of the design owner's tasks. This step has been involved in design protection and the 1999 Act does not change the situation.

Special Situation – More Than One International Classification for an Industrial Design

The reference in the 1999 Act, rule 7(7) to "class" evolved from a realization that some products are in rare situations classified in more than one international class. An explanation of this situation was given in the Diplomatic Conference preparatory document, H//DC/4 Prov., in reference to rule 16(2)(iv), page 38, footnote 13. This note stated, in essence, that WIPO would, in these rare cases, identify both classes in the International Registration. Assistance from a WIPO Locarno Agreement classification expert may be helpful in evaluating these situations.

International Registration – Effect

Meeting of Experts and Diplomatic Conference

A significant amount of time at the Meetings of Experts was spent on the International Registration effect. It is the heart of the Hague Agreement system. A thorough understanding of the International Registration effect is the key to applying the 1999 Act.

The Meeting of Experts discussions identified several aspects of International Registration effect that needed to be clarified. One point was when an International Registration was entitled to a filing date. Other issues were the effect on the national design law system as to infringement rights and prior art. These subjects, particularly the prior art one, remained an issue up to the final moments of the Diplomatic Conference.

The experts did not have a problem with the procedure for determining the filing date of an International Registration application. There was a need to clarify the International Registration effective date, the time when contracting parties had to treat the International Registration as filed. WIPO proposed a system that gave the International Registration the same effect as a national application on the effective date. The clarification was on what event created the effective date. There continued to be some confusion until the Diplomatic Conference on that point.

For some delegations the current status of U.S. law on prior art effect for design registrations was not satisfactory, and the pressure was on, as in other forums of international IP debate, to conform to a harmonized law approach on prior art effect. The U.S. and some non-governmental organizations at the Meetings of Experts strongly resisted this effort. At times it seemed that the treaty revision could not be finished, because of this issue. It was a tribute to the experts, Diplomatic Conference representatives, and officials of the conferences that progress was made and proposes were reached on most issues, enabling a Diplomatic Conference to be held. In the Diplomatic Conference these issues were reviewed and compromises made.

1999 Act International Registration Date

The procedural and legal importance of the International Registration date was evident from the discussions at the Meetings of Experts and Diplomatic Conference. The focus point was to clarify when it occurred.

The starting point was to establish the International Registration filing date, under various circumstances. If there was an irregularity, the date may be delayed [1999 Act, art. 9(3)]. When the International Registration application was filed indirectly, the International Registration date would be the date filed in the national office, if the application was received at the International Bureau within the prescribed time limit and it contained the mandated contents. A complete application directly filed with International Bureau would have a International Registration date when received by the Bureau.

The International Registration, when accepted, has the effect on the International Registration date of a national application filed in all designated contracting parties, no matter if it is published or deferred [1999 Act, art. 10(2)]. The International Bureau processing of an application is completed usually in two or three months. The publication of an International Registration under the 1999 Act will occur no earlier than six months from the International Registration date. If there is deferment, there is registration upon acceptance of the application and the International Bureau keeps the registration confidential.

The uniqueness of the deferment approach requires some historical review that is given in the Deferment Publication section. It was a compromise necessary for two quite different design protection systems to be within one treaty.

The result of these 1999 Act provisions is to have an International Registration effect that is the same as a national application filed on the International Registration date in the designated contracting party. This date will be about six months before the publication date of the registration. This background set the stage for some very serious discussions on the issues of the prior art effect of an International Registration in the U.S.

For contracting parties with a non-examining system, their national laws will determine the next step. The International Registration will be processed in the same way as a national design registration application to grant the design protection rights. For examination contracting parties, the International Bureau International Registration publication starts the examination process, under national law.

Provisional Infringement Rights

It was mentioned several times in the experts meetings that a contracting party may have provisional infringement rights that begin on publication of a design registration and are perfected upon the grant of rights [Sixth Session Draft Act notes #11.03]. It was clear from these discussions that the International Registration under the 1999 Act would have the benefit of provisional infringement rights, the same as national design registrations.

Infringement Rights Effective Date for International Registration

The 1999 Act required that an International Registration have at least the same effect in each designated contracting party as the respective nationally filed application [1999 Act, art. 14(1)]. The International Registration remains its own entity, but it follows the national office procedures.

The 1999 Act set up a limit on how long a contracting party could review an application to determine if it wanted to refuse it. This feature helped to prevent delays in registration. The refusal period ends automatically, unless otherwise provided, and the International Registration effect as a granted national right occurs automatically. The 1960 Act had this feature and it was used in the 1999 Act with some changes. In the 1999 Act, for non-examination offices the refusal maximum period was six months [1999 Act, art. 14(2) and rule 18(1)].

The International Registration infringement rights commenced based on national laws, and without a refusal no later than the end of that refusal period [1999 Act, art. 14(2)]. Some contracting parties have the grant of infringement rights on the filing date, so that the International Registration infringement rights would begin in these countries on the International Registration effective date. For examining offices there is a maximum refusal period of 12 months, if a declaration is filed, and when infringement rights commence can be specified in that declaration, within limits [1999 Act, rule 18(1)(b) and (c)].

The 1999 Act allows a variety of approaches on when infringement rights begin, to adjust to various national laws. There is clear pressure to commence the rights promptly.

International Registration Prior Art Effect

In principle the effect of an International Registration is the same as a national application and the resultant registration in each designated contracting party. Application of this principle to the U.S. created a major concern, as its law had for many years denied the effect as prior art for foreign filed applications filed under the Paris Convention, after the U.S. court decision of *In re Hilmer*, 359 F.2d 859, 149 USPQ 480 (CCPA 1996). This case denied the prior art effect of a utility patent under 35 U.S. sec. 102(e) to the filing date of a foreign application that properly claimed priority under the Paris Convention. It held that a foreign filed U.S. utility patent application is given prior art effect as of its U.S. filing date under 102(e) when the patent issues, even if there is a prior, foreign national application that provides a priority date under the Paris Convention. The same law applies to utility patents and design patents [35 U.S.C. sec. 171 (2004)].

When the Patent Cooperation Treaty (PCT) was negotiated, the U.S. required an exception to sec. 102(e) that filing of the U.S. designated PCT application had prior art effect under sec. 102(e) from the date when the PCT application was filed in the U.S., if it was in English.

It was apparent that the international community was unhappy with this situation, even before the Hague Agreement revision began. Efforts had been made for several years to change the U.S. law on prior art effect through international negotiations. It was argued that foreign inventors and designers were discriminated against. They filed applications in their country and then in the U.S., but their earlier, national applications properly filed under the Paris Convention in the U.S. did not result in prior art effect as of their national application filing date, even though it gave them the benefit of priority. They argued a level playing field was needed on prior art effect. This international campaign to change U.S. law now became part of the Hague Agreement revision debate.

At the Fourth Session of experts, the U.S. raised its concern that the Draft Act appeared to prevent the U.S. from applying the *In re Hilmer* law [Fourth Session Final Report #105 and #106], and others responded in support of the proposed text [#106-#110]. In the Fifth Session the U.S. proposed a provision called art. X, added to make an exception for the U.S. on this issue, so the *In re Hilmer* law could be applied. It was rejected by the most of the delegations at the Fifth Meeting of Experts [Fifth Session Final Report #117]. The Sixth Meeting of Experts Draft Act did not contain art. X. The stage was set for a major debate at the Diplomatic Conference on this issue.

There was some discussion at the Sixth Meeting of Experts that the U.S. was considering a change in the sec. 102(e) law [Sixth Session Final Report #94-#97]. This change was being discussed related to PCT, as a result of an article written by Lloyd Knight, a practitioner, and its impact was being felt in the Hague Agreement revision work [G. Lloyd Knight, *It's Time to Stop Discriminating Against PCT/USA National Phase Patents*, 79 Journal of the Patent and Trademark Office Society 385-400 (1997)]. It was up to the Diplomatic Conference on the Hague Agreement revision to resolve this debate for designs and probably for utility patents too, on an international level.

At the Diplomatic Conference, the opening statement from the ABA representative, William T. Fryer, III, made it clear that current U.S. law on prior art effect under the *Hilmer* decision was not compatible with the Draft Act, and this fact raised a serious question as to whether the U.S. would be able to accept the proposed treaty [see chapter I, General History of The 1999 Act, under the heading "Selected Opening Declarations"]. In the evolving discussions at the Diplomatic Conference the focus was on how to achieve a balanced solution. It was recognized that the design community was not the only U.S. group interested in this issue. Utility patents users had a big interest in the outcome. The design treaty revision was not the place for a full debate or comprehensive resolution of this issue.

The U.S. delegation agreed to follow the direction under consideration in the U.S. Congress for PCT. There was pending legislation to provide PCT related applications with prior art effect as of the PCT application date, under certain circumstances. The negotiations made it clear that the date an application was received by the International Bureau, or indirectly filed, was the International Registration date. There was no prior art effect back to a priority application filed under the Paris Convention on which the International Registration application was based. WIPO staff statements at the Diplomatic Conference confirmed that this interpretation was now clearly stated in the revised Draft Act [WIPO Records of the Diplomatic Conference ... Hague Agreement (Geneva) 496-497 (#938-# 944) (2002)].

This solution was accepted by the Diplomatic Conference, and represented a small step forward in international harmonization on prior art effect. Later, the U.S. Congress approved the PCT legislation, Public Law 107-273 on November 2, 2002, in the Intellectual Property and High Technology Technical Amendment Act, based on H. R. 2215 (107th Congress), with some specific requirements. It seems likely that the 1999 Act provision on prior art effect will be acceptable to the U.S. IP community and Congress with adjustment consistent with recent PCT related legislation.

One of the requirements that was applied to the recently enacted PCT prior art section 102(e) was that the PCT application had to be published in the PCT system and the application must be in English. It is likely that this requirement will be part of the 1999 Act implementing legislation in the U.S. Deferred International Registrations would have prior art effect, when they meet these requirements.

The argument against the U.S. requiring English text for a design patent published under the 1999 Act is based on the fact that it is not a requirement under 35 U.S.C. secs. 102(a), 102(b), or 102(d) for determining prior art. In addition, the International Registration is less dependent on text, since the drawing is the main claim for rights which can be interpreted usually without text help. Only in very unusual circumstances will the text portion of an International Registration be important in determining the scope of design rights. It is submitted that the prior art effect requirement of publication in English for an International Registration is not needed or justified for industrial designs.

Prior Art Effect of a Deferred International Registration

The other major issue regarding prior art effect was the legal effect of deferred International Registrations. The fact that that deferred, unpublished International Registrations would have the legal effect of a national application as of the International Registration date had to be taken into consideration. It is not a new question, at least under U.S. patent law, and there have been cases that analyzed the prior art issue, with specific reference to Hague Agreement based registrations [*see* Manual of Patent Examining Procedure (MPEP), U.S. Patent and Trademark Office (PTO), sec. 1504.02 (8th Edition, Rev. 1, Feb. 2003)]. Since the 1999 Act has new features, these prior cases and the MPEP will have to be reviewed to see if they remain applicable. The following brief review will consider only the U.S.

law, but it should be helpful to experts in each country that will have their own research to do on this topic.

The principle U.S. novelty provisions that should be considered to determine the prior art effect of an International Registration are 35 U.S.C. secs. 102(a), 102(b) and 102(d).

Sec. 102: A person shall be entitled to a patent unless:

(a) the invention was known or used by others in this country *patented* or described in a printed publication in this or a foreign country, before the invention thereof by the applicant for patent, or

(b) the invention was *patented* or described in a printed publication in this or a foreign country or in public use or on sale in this country, more than one year prior to the date of the application for patent in the United States, or

...

(d) the invention was first *patented* or caused to be *patented*, or was the subject of an inventor's certificate, by the applicant or his legal representatives or assigns in a foreign country prior to the date of the application for patent in this country on an application for patent or inventor's certificate filed more than twelve months before the filing of the application in the United States

...

[Emphasis added].

The fact that a published International Registration is prior art under 102(a) and (b) on the publication date is clear. It is accessible to the public. The next question is when an International Registration is "patented" under secs. 102(a) and (b). These issues can be debated, as evidenced from *In re Carlson*, 983 F.2d 1032, 25 USPQ 2d 1207 (Fed. Cir. 1992), where public access was the primary factor. A deferred International Registration is kept confidential, so there is no public access. The question of whether there is prior art effect under secs. 102(a) or 102(b) from a patent with its text not yet available to the public needs to be answered. The law developed around sec. 102(d) may be helpful in making this analysis. On the other hand, the overall role of sec. 102(a) is to deny a patent if the public has access to the invention of another person. This principle suggests that access should be required to the patent that is prior art under sec. 102(a). In contrast, sec. 102(b) involves a personal choice for the inventor, similar to sec. 102(d). The principle of sec. 102(b) for patented supports the same interpretation as under sec. 102(d), discussed below.

Under sec. 102(d), grant of a patent anywhere in the world by an inventor can be prior art against an application filed in the U.S. by the same inventor, under certain circumstances. The patent must have been filed in that foreign office more than six months before the U.S. application. The foreign patent must have been granted and the enforcement rights obtained before the filing of the U.S. application. With the timing considerations met, the issue boils down to when the patent rights are available. The fact that the rights may go back when granted is not the critical point. The time when the rights were awarded, even if they go back to an earlier time, is the key date for determining the prior art effect date for a foreign patent.

There are several U.S. cases that have addressed the question of when there is prior art under sec. 102(d). The two most recent primary authority decisions were by the Federal Circuit, Court Appeals. The abovementioned *In re Carlson* case had *dicta* regarding the law under sec. 102(d) The Federal Circuit reviewed the case law on 102(d) for a utility patent, in the decision of *In re Kathawala*, 9 F.3d 942, 28 USPQ 2d 1785 (Fed. Cir. 1993) and discussed several related design patent cases. It should be noted that sec. 102(d) applies to design and utility patents. The court held that the critical fact in Kathawala was that the patent rights had been granted and the fact that there was no opportunity for public access at that time to the patent text did not matter.

The Kathawala decision based its holding on several prior cases. The issue of design registration secrecy was addressed in a Court of Customs and Patent Appeals decision, *In re Monk*, 588 F.2d 308, 200 USPQ 129 (CCPA 1978). A UK design registration had been published and the issue was when the patenting occurred under 102(d). Since the rights were not awarded until later, when the patent was sealed, their was no patented design under sec. 102(d) until the later date. The *Monk* court decision reviewed the legislative history of sec. 102(d) in coming to this conclusion. The original line of cases on sec. 102(d) were the decisions of *In re Talbott*, 443 F.2d 1397, 170 USPQ (CCPA 1971) and *Ex parte Weiss*, 159 USPQ 122 (Pat. Off. .Bd. App. 1967). These cases dealt with design registrations that were kept secret and, in spite of this fact, a patented design was found under sec.102(d). Other PTO administrative decisions addressed design registrations obtained under the Hague Agreement. These cases carefully analyzed the Hague Agreement system and procedure in each country where the filing occurred to determine if the design was patented: *Ex parte Marinissen*, 155 USPQ 528 (Pat. Off. Bd. App. 1966), *Ex parte Appeal No. 239-48* (Pat. Off. .Bd. App. 1965). Other early cases on this subject were *Ex parte Einfalt*, 155 USPQ 529 (Pat. Off. .Bd. App. 1966) (German design registration was held in secret and found patented.) and *Ex parte Lancaster and Stein*, 151 USPQ 713 (Pat. Off. .Bd. App. 1965) (Canadian design registration had rights granted and was patented).

In applying these principles to deferred International Registrations, it will be a matter of analyzing the national law applicable to the International Registration. The major task is to find out when the design rights were first available. An examining contracting party has the right to set the time when an International Registration takes effect [1999 Act art. 14(3) and rule 18(1)(c)(ii)]. The International Registration effect for non-examining contracting parties would at the end of the six months refusal period, automatically, as the filing date in the designated countries [1999 Act, art. 14(2)]. A contracting party could mandate the International Registration rights to occur effective on the International Registration date, the effective filing date in the national system. The rights grant for examining offices would likely be set to coincide with completion of the administrative work on the International Registration.

A variety of situations could be encountered, and each one would have to be analyzed to determine when sec. 102(d) would apply. The 102(d) main principle is that there is no prior art effect from a foreign patent if the date limitations in sec. 102(d) are not satisfied, and if the patent right was not yet enforceable at the critical time. Another principle to apply is that a granted patent right that is enforceable is all that is needed. There does not have to be a publication of the patent, or access to the patent document.

This example using the more complex U.S. design patent system should help demonstrate some of the considerations that may be involved in other design registration systems. It helps emphasize the need for designated examining offices to request receipt of a deferred International Registration copy, as permitted under the 1999 Act [1999 Act, art. 10(5)]. These copies can be used in the national office search files and kept available for use in post-grant litigation or administrative proceedings.

The U.S. is in the unique position that it will not receive copies of the deferred International Registrations if it decides not to allow a deferral period. Copies go to contracting parties designated in the International Registration. There will be no deferral of the U.S. designated International Registration under this situation. In that case, some prior art not available to the U.S. office may exist on International Registrations it will be examining. This situation should be corrected by allowing any examining contracting party office, whether or not designated in an International Registration, to receive copies of deferred International Registrations.

Before the recommended change, it is not much different than the current situation under U.S. law. Foreign design registration systems are now deferring registrations and granting enforceable rights during deferment without the U.S. receiving copies of these documents. It is a situation that can be improved in due course by obtaining access to deferred International Registrations. Whether there is a sufficient number of likely disputes to warrant the change is another question. The administrative handling of these deferred applications is another concern. As an alternative, it may be less of an administrative burden to require a foreign International Registration owner to disclose to the office during examination information on earlier filed foreign applications on the design. In litigation, the same information would be important.

Multiple Designs International Registration

Meetings of Experts and Diplomatic Conference

The Hague Agreement Act of 1960 permitted a maximum of 100 designs in an International Registration when they were in one international class. It was a way to reduce costs, as the extra designs had a substantially lower filing fee. There was strong support at the Third Meeting of Experts for adopting the 1960 Act procedures on multiple designs in the Draft Act [Third Session Final Report #81]. At that time, however, some examining countries expressed a concern because of the unity of design concept that allowed only one design as a registration. The compromise was that multiple International Registrations could be file in any contracting party, but the registration would be divided into a separate application for each design concept. This proposal left the examining offices free to continue their existing laws, a goal of the revision whenever possible.

The fee for filing a multiple design International Registration application would remain low, but in each contracting party the fee for each separate design would be the full, initial fee for one design. It was clear that the earlier Hague Agreement users were comfortable with the administration of the same approach used in the 1960 Act. Other experts recognized the cost savings and efficiency of the multiple designs International Registration. There was some discussion at the Diplomatic Conference on whether the number of designs in an International Registration application should be limited or expanded. The Diplomatic Conference accepted the experts' proposal of a maximum of 100 designs from the same International Class in an International Registration application, as a necessary flexibility to accommodate non-examining and examining systems [WIPO Records of the Diplomatic Conference ... Hague Agreement (Geneva) 409 (#146-#149) (2002)].

1999 Act Multiple Design International Registration

A maximum of 100 designs from one International Class can be in an International Registration application [1999 Act, art. 5(4), and rules 7(3)(v) and 7(7)]. The design reproductions must be placed on sheets in a prescribed format [1999 Act, Administrative Instruction 405(a); *see* Appendix X for example from WIPO Designs Bulletin].

The 1999 Act recognized that a unity of design concept could be used for an International Registration by an examining contracting party, if it made a declaration to that effect [1999 Act, art 13(1)]. The right of an examining office to refuse an International Registration with more then one design concept was recognized, but the right of the applicant to retain their filing benefit for each design concept was preserved. The International Bureau would keep appropriate records on new proceedings due to a division of a multiple design International Registration [1999 Act, rule 18(3)].

The sections on Reproductions and Cost to Users have useful background information related to multiple designs International Registrations.

Postal Delays

1999 Act Procedure for Postal Delays

In the practice of law, due dates and how to comply with them by sending documents to be received by a certain date are very important matters. The Meetings of Experts often referred to these concerns. A set of precise limits on when to file documents for WIPO to receive them within a required date was provided [1999 Act, rule 8]. These procedures may be updated frequently through the Administrative Instructions [*See*, for example, Administration Instruction 501].

Specific situations were described in rule 8, and it was a matter of proving the facts to see if the requirements were met. All records on mailing steps taken and receipts from the mailing office need to be retained, in case a question arises due to postal delays.

Priority and Filing Strategy

Paris Convention Applicable to the 1999 Act

An International Registration can be based on a national application to receive benefit of the International Registration filing date, under certain circumstances [1999 Act art. 6(1)]. The priority application must have been filed in a country that was a party to the Paris Convention or the WTO and filed within the six months Conventional period from the first filed application for the industrial designs.

In the same respect an International Registration can be the basis for priority under the Paris Convention for a national application in a member of the Paris Convention [1999 Act, art. 6(2)].

An extensive discussion occurred at the Diplomatic Conference to clarify the terms: filing date, International Registration date, and priority date, that appeared in the Draft Act. The effect of an irregular International Registration application with a later International Registration date from the filing date was one situation considered. At the conclusion of the Diplomatic Conference these terms were clearly defined [WIPO Records of the Diplomatic Conference ... Hague Agreement (Geneva) 410-413, 429-430, 435, 454 and 455 (#157-#183, #313-#317, #369, #528 and 529) (2002)]. The International Registration priority filing date and the International Registration date are the same date, when a complete application is received. The International Registration filing date for priority purposes of an irregular International Registration application will be the date when the application was filed. The International Registration date for that application will be when the application was completed.

Strategy for Using the Paris Convention and 1999 Act

The Paris Convention and the 1999 Act create options for obtaining design protection in contracting parties under the 1999 Act and in other Paris Convention member countries. The short six-month Paris Convention period does not give much time to make decisions on where to obtain foreign protection. If the direct national route is used, all applications must be filed and national office fees paid within the six months after the corresponding national application was filed. Filing an International Registration application with priority to a national application under the Paris Convention creates the same six months pressure.

If deferment is available, the final decision on whether to complete the International Registration for all or some of the designs or designated countries can be delayed up to 30 months, depending on the countries designated and their design registration deferment laws. If an International Registration is based on a Convention application, an examining national office filing would proceed and give an indication on the likelihood of significant protection in the home market where usually the product will offered for sale first. It will be acted on sooner than if it was processed as part of the International Registration. These are factors to consider on whether to file a national application first, or rely on the International Registration for filing in applicant's contracting party. The benefits appear to favor filing a national application first, at least for the U.S. design owner.

Looking at the filing from a U.S. design owner's perspective, for example, a national application and an International Registration under the Paris Convention based on the U.S. filed application would set up a situation where deferment may be more likely in many contracting parties. If the U.S. does not allow any deferment, which is likely to be the case, an International Registration designating the U.S. for protection could not have any deferment. If an International Application not designating the U.S. is filed, based under the Paris Convention on a prior filed U.S. application, deferment may be available on the International Registration application. The deferment makes available several cost saving steps.

The same general approach would be useful to all parties filing in the U.S. or other countries where the same limiting effect on deferment existed. A separate national application or International Registration application would be filed requesting no deferment for contracting parties not allowing deferment. An International Registration application would be filed for contracting parties that allow deferment, to provide the flexibility in cost savings. The multiple designs in an application, where appropriate, would add to the flexibility and cost savings.

Product Indication

Meeting of Experts and Diplomatic Conference

One of the concerns expressed at the Meetings of Experts was the impact of the requirement to indicate the product that is the design or on which the protected design was used. This requirement was proposed as a mandatory item for obtaining a filing date.

In design practice for some national systems it is very important not to limit by description the design scope. The title of the design, or a required description may be interpreted as limiting the products on which the design may be used. If the national design law protects a registered design for any type of product on which it is used, there is no problem as far as protection scope is concerned with how the product is indicated in the application.

The recommendation of experts was to allow flexibility in preparation of the International Registration application product indication. Practitioners would develop the right product indication statement strategy for the designated countries. The Draft Act text on this point did not change significantly after these discussions, but there were some assurances made [Fifth Session Final Report #41]. WIPO assured the experts that its role would be only to check that a product indication was in the application and not to become involved in selecting the wording. This comment is important, as discussed below in connection with preparation of the International Registration application.

1999 Act Requirement for Product Indication Identification and Strategy

The 1999 Act required the product indication statement to obtain a filing date [1999 Act, art. 5(1)(iv)]. It is suggested in rule 7(3)(iv) that the product indication use terms from the International Classification to simplify the determination of product classification.

If the International Classification terms are broadly stated, it may be a wise step to follow the suggestion to use International Classification terminology. In other situations it may be a useful practice to add scope to the product indication statement for more effective protection in some designated contracting parties. The approach would be to indicate more than one name for the product type or more than one type of product that incorporates a design. Rule 7(3)(iv) already suggests flexibility in its reference to the "product or products" that may incorporate the design.

1999 Act International Application Form

The International Registration application form has several suggestions written on it to help fill out the product indication part of the form [Appendix VIII]. In #8 several details are entered about the design. The "products" which make up the design are named. This information is a list of the products for which reproductions or specimens are included. The International Classification subclass for each product is stated, along with the number of designs submitted in the application. Each design must be in the same International Class. The product indication would be the identification entered under the product(s) heading in #8.

The application form recommends in #8: "[i]f the industrial design consists of a product, indicate the usual generic name, for example: 'chair'. If the industrial design consists of a decorative motif, indicate the product(s) in relation to which it is to be used, for example: 'drawing to be used with 'dishes' or 'motif for textiles". These suggestions are consistent with the aforementioned goal of a broad product indication statement to provide an adequate scope of protection needed for some national systems.

Publication

Meeting of Experts and Diplomatic Conference

Another subject that had a lot of discussion at the Meetings of Experts was the International Registration publication. The focus was on the International Bureau publication stage and whether a contracting party could publish the International Registration after the International Bureau published it, and if so, under what circumstances.

There was discussion at the Meeting of Experts on a two-stage International Bureau publication approach. In this proposal, non-confidential information would be published on registration, and after the deferment period ended the reproduction would be published. This proposal was rejected in the Sixth Meeting of Experts, primarily because of expense to users [Sixth Session, Draft Act note 9.11]. The experts agreed that no information would be published until the deferment period ended.

A subject that did not receive much discussion by the experts was when the International Bureau would publish an International Registration after formal processing was completed. Under the 1960 Act usually the publication took place within three months of the application receipt by the International Bureau, after the contracting parties accepted the International Registration. There was a six-month period allowed for all countries to refuse the effect of an International Registration, so final approval to publish had to wait for that period to end, unless approval was received earlier.

WIPO added a provision to the Draft Act that no publication would occur sooner than six months from the design, unless the design owner requested it. The discussion was that some delegations wanted this provision to ensure there would be a minimum mandatory deferment period for most International Registrations.

The argument was that a minimum deferment served a useful purpose, to prevent early access to product developments. For contracting parties that had no deferment, this six-month provision gave some balance on this important issue. The built-in six-month delay was the result of difficult negotiations on several issues during the Sixth and Seventh Meetings of Experts (see General History chapter for more details) [Seventh Session Draft Act note 9.03]. An International Registration owner could request earlier publication. The Diplomatic Conference accepted the experts' approach to WIPO International Registration publication, but it remain for the Diplomatic Conference to resolve the issue of contracting party publication, discussed below.

The Meeting of Experts and Diplomatic Conference had considerable discussion of the role of contracting party publication. What evolved was a realization that national information on an International Registration in the national language was considered necessary by some experts. Also, in examining system countries there needed to be communication of the final form of the protection rights. The International Bureau was not commissioned to republish an amended International Registration.

A consensus was reached that publication of an amended International Registration could be made by a contracting party for a fee to the owner, but considerable uncertainty about the scope of this agreement remained [Seventh Session Final Report #47-#52]. This exception was considered within the original principle that republication of the International Registration Bulletin in original or translated form could not be charged to the International Registration owner.

At the Diplomatic Conference, after considerable discussion, it was agreed that contracting party republication of an International Registration could not occur at the registration owners's expense [WIPO Records of the Diplomatic Conference ... Hague Agreement (Geneva) 469-470 (#677-#689) (2002)]. The expert's proposal that an exception should exist was rejected. The Draft Act did not specifically prohibit a contracting party from charging a design owner for publication of an amended International Registration. One of the reasons given for resolving this issue, eliminating the exception, was the relatively low cost of republication using the Internet. The delegates seem to conclude it was not a significant cost issue [WIPO Records of the Diplomatic Conference ... Hague Agreement (Geneva) 469-470 (#677-689) (2002)].

1999 Act International Bureau and Contracting Party Publication

When the International Bureau completes it review and the International Registration is ready for publication it will be published no earlier than six months from the date of the International Registration, unless requested by the design owner [1999 Act, art. 10(3)(b) and rule 17]. A deferred International Registration would be registered promptly and confidential copies sent to designated contracting parties

[WIPO Records of the Diplomatic Conference ... Hague Agreement (Geneva) 441 (#413 and #414) (2002)]. Usually the process takes no more than three months. There is quicker registration of deferred International Registrations.

A copy of the registration is sent to each designated contracting party using a Bulletin published on the WIPO website [1999 Act, rule 26(3)]. This publication has effect for all contracting parties and no separate publication by contracting parties can be made at the Industrial Registration owner's expense [1999 Act, art. 10(3)].

Recording Ownership and Other Information

Meeting of Experts and Diplomatic Conference

One major concern for the U.S., and perhaps other countries, was the recording of information on an International Registration [Seventh Session Final Report #79 and #80]. The U.S. representative pointed out that its design law included a legal effect from recording ownership [35 U.S.C. Sec. 261 (2004)]. The concern was over having two databases involved in determining legal rights. The Diplomatic Conference resolved this issue by allowing a contracting party to declare that changes in ownership would not have legal effect until the ownership documents are received by the contracting party [WIPO Records of the Diplomatic Conference ... Hague Agreement (Geneva) 482-484, 498-499 (#809, #829, #952-#963) (2002)].

1999 Act Recording Procedures Ownership and Other Information

The International Bureau provides an important service as a central source for information on International Registrations. It publishes the International Registration information initially [1999 Act, rule 17], and if national office records change the International Bureau publishes the changes [1999 Act, art. 10(4)]. This up-to-date information is available on the WIPO website.

The effect of a recording of information in the International Bureau is the same as if made in the national office, with the exception of the provision that a contracting party may declare that the ownership documents be received by its national office before the legal effect begins [1999 Act, art. 16(2)]. In addition, a further limitation on ownership transfer was that the transferee must be eligible to own this International Registration, as discussed below.

1999 Act Ownership Transferee Entitlement

Since the 1999 Act is a closed system, only a party entitled to file may participate. Whether an applicant can designate a contracting party for protection depends on whether the applicant's contracting party has a common treaty relation in the

Hague Union with that contracting party and which treaties are involved. [1999 Act, art. 11]. The same entitlement requirement applies to transfer of ownership. For example, an International Registration owner whose contracting party is a member of the 1999 Act can transfer ownership of the International Registration to another party entitled to file under the 1999 Act. As another example, an owner of an International Registration under the 1934 Act cannot transfer an International Registration to a party that is entitled only to file under the 1999 Act. If the latter party was also a member of the 1934 Act, the transfer would be permitted.

Another unique feature of the 1999 Act is that its fee structure helps in the transition of ownership transfers. An International Registration can be renewed for all contracting parties and designs irrespective that ownership transfer of part of the International Registration has not been completed [1999 Act, art 17(2) and rule 24(1)]. When the situation permits transfer to an appropriate owner the renewed International Registration is available for that transfer.

Refusal of International Registration

Meeting of Experts and Diplomatic Conference

One of the main dramas of the Meetings of Experts and the Diplomatic Conference was on how the issues surrounding the national office International Registration refusal would be handled. The negotiation went to the brink of failure and pulled back. The General History chapter has a more complete review of these events.

The 1960 Act of the Hague Agreement gave a national office a maximum time of six months to refuse the effect of an application. No further review of the International Registration after the initial refusal could occur. The effect of no response was automatic approval. It was not a workable system for most countries with examining offices.

Several procedures had to be revised by the experts to find an approach acceptable to the all examination offices, and to the delegations with non-examining offices. The time period for refusal had to be extended. A unique feature of the U.S. design patent system that allowed further novelty rejections had to be accepted. Finally, the line between what could be refused by an examining office and what the International Bureau could approve and not be reevaluated by examining offices had to be drawn. All these tasks were accomplished by the experts under the guidance of WIPO staff in a series of dramatic steps. Each of these topics will be dealt with under separate headings and the events leading up to the 1999 Act and its provisions will be reviewed in that section.

At the Diplomatic Conference the UK requested clarification of the term "examining office", a term defined in 1999 Act, art.1(xvii). The UK office planned to examine International Registration applications for compliance with the UK law must fit and must match exceptions to registration. This type of examination did not come within the scope of the Draft Act definition of "examining office". The WIPO staff explained that "examining office" was a key concept for several provisions in the Draft Act, based on an office performing a novelty examination. Special treatment was given for this type of system. It was pointed out that non-examination offices, such as the UK, could refuse an International Registration application during the six-month period provided in the Draft Act for refusal [WIPO Records of the Diplomatic Conference ... Hague Agreement (Geneva) 402 (#82-#85) (2002)]. The WIPO view on this definition of "examining office" was consistently applied in the Diplomatic Conference.

Refusal Period Extension

The only way to resolve the conflict with the six-month refusal period of the 1960 Act and examining office procedures was to extend the refusal time. Initially, the experts discussed a maximum 30-month refusal period. This time period appeared to be acceptable to the U.S. and Japanese government representatives. As the negotiations on this point and others became more intense the time period was reduced to 18 months, and the U.S. Government representatives indicated a 12-month period would be acceptable, based on the performance of the U.S. design patent examination process at that time, and if the grounds of refusal issue was resolved satisfactorily.

The 1999 Act continued the six-month refusal period for non-examination contracting parties [1999 Act, rule 18(1)(a)]. The contracting parties with examining offices were given the right to obtain, by declaration, an extra six months, for a 12 months total to refuse an International Registration [1999 Act, rule 18(1)(b)].

The 12 months refusal maximum was so important to the negotiations, that the 1999 Act rule provision setting that limit could only be changed by unanimous vote of the Assembly [1999 Act, rule 33(1)(ii)].

Grounds for Refusal

It became clear that for some examination offices there was a basic problem with allowing one opportunity to challenge an International Registration. In these systems it was important to add new grounds of rejections where prior art was found after the first refusal. Offices with these systems consider it very important that intellectual property rights should be granted with a reasonable novelty search and if newly discovered prior art was discovered there should be a new rejection. Several times during the experts meetings, and during the Diplomatic Conference, it was agreed that the required refusal was a report of what was known at that time, equivalent to a U.S. first office action [*see*, for example, Sixth Session Draft Act note 10.5 and Final Report #49; [WIPO Records of the Diplomatic Conference ... Hague

Agreement (Geneva) 435 (#365-#368) (2002)]. The office could add other refusal grounds later, while an International Registration application was pending.

Examining Office Failure to Report Refusal Due to Misplaced Records

The U.S. delegation at the Diplomatic Conference continued to express concern over the automatic rights grant in a contracting party if the examining office did not report a refusal within the appropriate time period. It needed assurance that misplaced office records or other errors that resulted in failure to communicate the refusal would not result in an inadvertent grant of International Registration rights. It was agreed at the Diplomatic Conference that an examination country could report the refusal after the error was discovered and the rights grant would not occur [1999 Act, 18(1)(c)(ii); [WIPO Records of the Diplomatic Conference ... Hague Agreement (Geneva) 493-495 (#904-922) (2002)]. The U.S. willingness to reduce the refusal time maximum to 12 months clearly encouraged the cooperation of the other delegations in this negotiation.

International Bureau Role in Approving International Registration Formalities

It took some time to convince the experts what the role of the International Bureau was. The formality part of the International Registration review was to be handled by the International Bureau. The decisions on whether an applicant was entitled to a filing date, or other related formalities, were for the Bureau to make. The 1999 Act made this point clear [1999 Act, art. 12(1)].

The dividing line between the national office and the International Bureau roles was best illustrated by reference to the review of the International Registration reproductions. The Bureau prepared the application file and other record information, followed by a review of the application form to see that all required information was present. The reproductions of the design were examined to determine that at least one reproduction, or specimen, if appropriate, was present and the format was correct. An examining office could review the reproduction again on substantive grounds to see that the design was adequately disclosed. The line was drawn between formality handled by the Bureau and substantive law issues in the province of the examination offices. Another example of the examining office role would be the determination of whether a reproduction had new matter in relation to a specimen that it replaced [1999 Act, rule 7(6)].

Conclusions

This part of the 1999 Act review deserves a separate conclusion and tribute to the experts, government delegates and WIPO staff that steered the negotiations to an approved treaty. These additions, adjustments and interpretations were major steps in achieving success in completing the 1999 Act, so that examination offices could effectively interface with it.

Regulations

General

The 1999 Act regulations have been approved as the common regulations for the Hague Union, governing the 1934, 1960 and 1999 Acts [1999 Act, art. 24]. The regulations govern the details of the Geneva Act (1999) implementation. Subject to important exceptions requiring a high vote, the regulations can be changed by two-thirds of the votes cast where a quorum is present [1999 Act, art. 21(5)]. The latest version of the common regulations were effective on April 1, 2004, when the 1999 Act became operational.

A number of key rules in the common regulations could not be changed without a higher vote of the Assembly [1999 Act, rule 33(1)]. The two rules requiring unanimous vote were the indirect filing national security six-month review period [1999 Act, rule 13(4)], and the International Registration refusal extension of six months [1999 Act, rule 18(1)]. Another set of provisions required a vote of four-fifths of the votes case, with a quorum present. These provisions were the limit of 100 designs of the same International Class in an International Registration [1999 Act, rule 7(7)], the extent of a contracting party examining office review [1999 Act, rule 9(3)(b)], the 30-month maximum deferment [1999 Act, rule 16(1)(a)], and the mandatory six months before the International Bureau could publish an International Registration [1999 Act, rule 17(1)(iii)]. The special voting conditions made clear which topics were most important in the 1999 Act negotiations.

Renewal

Meeting of Experts and Diplomatic Conference

Fees were the primary issue that was discussed on renewal of an International Registration at the Meetings of Experts. The experts were very receptive to meeting the U.S. special situation where payment was for the entire term of 14 years protection upon acceptance of the application. The U.S. requested the flexibility to include in the individual designation fee the full term renewal fee, and a suitable provision was included to allow this result [1999 Act, art. 7(2)]. The Diplomatic Conference accepted this approach.

1999 Act Renewal Procedures and Strategy

Renewal of an International Registration is flexible and arranged to help a user eliminate design protection that is no longer needed [1999 Act, art. 17(2) and rule 24]. The International Bureau will publish renewal information and the status of an International Registration can be checked on the WIPO website [1999 Act, rule 26(1)(vi)]. The Bureau will send out an unofficial notice of renewal [Administrative Instructions sec. 701, but the provision cautions the owner that the fact the notice is not received is no basis for avoiding abandonment [1999 Act, rule 23]. There is a six-month post-due date renewal procedure requiring a surcharge [1999 Act, rule 24(1)(c)].

As mentioned in the Meeting of Experts summary, there was considerable flexibility allowed in the Draft Act for a contracting party to arrange the renewal fee payment procedure. Whether a designation fee was paid in two parts, for example, was up to the contracting party's office to arrange and notify the International Bureau of the plan. It is likely that the same fee procedure applied to national applications will be used for international applications, giving the contracting parties a way to easily interface with the 1999 Act.

Renewal periods are for five-year intervals, unless changed by a contracting party's declaration. While there is a 15-year minimum term from International Registration date, the protection term can go to the end of the term for each contracting party under national law.

Flexibility is built into the 1999 Act to renew as to specific designs and for selected contracting parties, in any combination, to allow a owner to decide what designs and where protection should be renewed. This approach may allow significant cost savings.

Special Renewal Situations

International Registrations can be renewed even though there has been a refusal recorded [1999 Act, rule 24(2)(c)]. This step is needed to preserve the status if an appeal or litigation is under way, for example.

Another unique situation is that an International Registration with its term ended, under national law, can still be renewed, to preserve its status in a controversy, for example [1999 Act, rule 24(2)(b)]. International Registrations that have had an invalidation recorded cannot be renewed for that contracting party [1999 Act, rule 24(2)(d)].

When a transfer cannot be recorded due to a lack of entitlement, a renewal can keep that International Registration active and available to complete the transfer at the appropriate time.

Renewal Amount

The International Bureau has a basic fee to cover its costs for renewal [1999 rule 24(1)(i)]. A standard designation renewal fee may be accepted by a contracting party, or an individual designation renewal fee can be set by declaration from a contracting party [1999 Act, art. 7(2)].

Representation

Meetings of Experts

It has been the policy in prior WIPO-administered treaties to not require qualifications to represent a party before the International Bureau, and this policy was followed again in the Draft Act. It is up to the design owner to find a person with the necessary qualification to handle a matter before the Bureau. In the national offices, the policy has been to let the contracting party decide the representation qualifications. [WIPO Records of the Diplomatic Conference … Hague Agreement (Geneva) 474-476 (#720-744) (2002)].

The subject of representation before the International Bureau and national offices came up several times at the Meetings of Experts [*see*, for example, Fourth Session, Final Report #117 and Seventh Session Final Report #114]. The experts were satisfied with the procedure set up by the Draft Act, and it was accepted at the Diplomatic Conference.

1999 Geneva Act Representation before the International Bureau

It is important to understand that an International Registration can be filed by the design owner or its representative. Any person that design owner selects can be the representative. There are no International Bureau qualifications for representation. There are specific procedures for appointment and removal [1999 Act, rule 3].

National Office Representation

Once an International Registration has been transmitted to a contracting party, and there is a matter to resolve in the national office, it is up to the national office whether a representative requires certain qualifications. It is likely the current practice for a national application will be followed by a contracting party.

Reproductions (Photos and Graphic Representations)

Meeting of Experts

In general, the Meetings of Experts accepted the successfully applied procedures on design reproductions developed for the Hague Agreement 1960 Act.

An issue that was discussed at some length was the problem created by filing a specimen and then replacing it with a reproduction. The question was who would review the reproduction to see if was the same design as the specimen? A related concern was who will have access to the files to make the review? The conclusion reached was that it was the role of national offices to make this determination.

Other topics discussed at the experts' meetings were national system requirements for graphic representations and use of photos. A good example and a positive result concerned the U.S. design patent system requirement that shading lines had to be used. In a step that showed interest in the Draft Act work and cooperation, the U.S. delegation announced removal of the U.S. requirement for appropriate shading in all design patent drawing [Seventh Session, Final Report, #12]. U.S. rules now merely suggested use of shading. The experts recognized a need for harmonization of reproduction procedures, but the Draft Act preparation was not the right time to complete that work. Ii would be easier to do it after the treaty had been operational for a while.

1999 Act Provisions for Use of Reproductions and Specimens

Each International Registration has a reproduction of the design for which protection is being sought. This visual presentation can be in either or both a photo and graphic representation. For two-dimensional designs only, a specimen of the design can be filed initially to obtain a filing date. The reproduction must clearly depict the design, and usually for product shapes at least one three-dimensional view was needed to accomplish that goal. National design registration systems differ significantly in what type of reproductions are allowed and how to present the design.

For these reasons, the 1999 Act is specific on some aspects of reproduction preparation for the International Registration application. It does not create a uniform approach for all national offices to use, but that goal would be worthy of future effort.

Basic International Registration Application Reproduction Principles

1. Each International Registration must contain a reproduction to obtain a filing date [1999 Act, art. 5(1)(iii)].
2. The reproduction may be in the form of a photo, graphic representation, or initially for two-dimensional designs, a specimen of the product showing the design [1999 Act, art. 5(1)(iii), rule 9(1)].
3. Color or black and white photos and graphic representations can be used [1999 Act, rule 9(1)].
4. Color or black and while photos and graphic representations may be used together in an International Registration [1999 Act, Administrative Instructions 401(a)].
5. More than one view may be shown of the product with the design [1999 Act, rule 9(1)].

6. A reproduction should show clearly all design features forming the design. This careful preparation will insure that the desired legal effect in the designated contracting parties where protection is sought.
7. Portions of a design reproduction may be excluded from protection, by using dotted or broken lines, or by disclaiming these features in the International Registration written description [1999 Act, Administrative Instructions, sec. 403]. Preferably, this disclaimer can be confirmed in the application text by stating that the purpose of the dotted or broken lines is to identify features that are not protected. Since the International Registration will go to contracting parties with varying procedures, this written confirmation will help inform the national office and the public on what is protected. The 1999 Act authorizes use of broken and dotted lines for this purpose, it is expected that national offices will provide the appropriate regulation authorizing their use and emplaning their legal effect. The 1999 Act has accomplished another small step in procedural harmonization for reproductions and the interpretation of International Regulation reproductions. More work remains to be accomplished.
8. Refusal of a reproduction on formality grounds, because it does not have the same design as a prior filed specimen is not the job of the International Bureau. This issue can be raised in the national office as a substantive law issue of whether new matter is present. The procedure is not in conflict with the prohibition in the Act for national offices reviewing formal matters approved by the International Bureau [1999 Act, art. 12(1)].
9. Shading or other graphic technique are not required, but they are good steps to use in showing a design's features clearly [1999 Act, administrative Instructions 404(b)].

Summary of Basic Design Reproduction Format Requirement

1. Single Design International Registration
   (a) A two-dimensional design can have more than one reproduction [1999 Act, rule 9(3)(b)].
   (b) A three-dimensional design can have several views and it is up to the applicant to select them, in terms of number of views and the perspectives. Usually, a view from each side is used (six views) plus at least one perspective view [1999 Act, rule 9(3)(b)].
   (c) The layout on a page of the different views for a simple design requires following precise dimensions and arrangement [1999 Act, Administrative Instructions, sec. 401]. As shown in Appendix X, the views are numbered

to identify the design and view [1999 Act, Administrative Instructions, sec. 405(a)].

(d) Each reproduction must have a plain background, so the design stands out alone [1999 Act, Administrative Instructions sec. 402(a)].

(e) Since Industrial Registration publication costs are priced by the reproduction page, it may be possible, depending on the reproduction size, to have several reproductions on a page. No more than 25 reproductions can be on a page [1999 Act, sec. 401(c)].

2. Multiple Design International Registration

(a) The abovementioned requirements for a single design International Registration apply to a multiple design application. The reproduction number changes for each separate design [1999 Act, administration instructions, sec. 405(a)]. The WIPO Bulletin International Registration publication shown in Appendix X illustrates the arrangement for multiple design reproductions.

3. Photos and Graphic Representations – General Comments

(a) The photos cannot be retouched [1999 Act, Administrative Instructions, sec. 404(a)], eliminating use of photo editing to disclaim features. Graphic representations offer a better way to select what is the protected design, including disclaimer by using broken or dotted lines for features that are not protected.

4. Specimens.

The procedures for specimens are specific and covered in the section on that subject.

Resources for Information on Reproduction Preparation

The 1999 Act rules and Administrative Instructions contain precise reproduction preparation details, based on prior International Registration practice under the Hague Agreement. A sample of the International Bureau Designs Bulletin is shown in Appendix X. This document shows the general format for placing and number the reproductions. A review of International Registrations, available on the WIPO website under the Hague System is another useful way to see how a particular type of design has been depicted and arranged.

Specimens

Meeting of Experts and Diplomatic Conference

A specimen is a sample of the original two-dimensional design, like a piece of fabric with a design on it. The experts' favored use of specimens to establish an International Registration filing date is only for two-dimensional design [Second Session Final Report #54- #57]. Expanding the use to three-dimensional products would create storage problems that would be hard to manage.

The concept proposed for use of specimens was to combine it with the deferment practice. If deferment of a two-dimensional design was requested, a specimen could be used in the International Registration application to obtain a filing date, instead of a reproduction (photo or graphic representation). Near the end of the deferment period, a reproduction of the design would be substituted and the procedures would be the same for all International Registrations from then on.

Some experts spoke in favor of using specimens for deferred International Registrations as a way to make the Hague Agreement more attractive to the textile, fashion and other industries that have large numbers of designs needing protection. The procedure would allow a lower initial cost for filing and other savings later. The impact of the textile and fashion industries on the expert meetings and the Diplomatic Conference was significant. Their efforts were supported by an international agreement, TRIPS, that stated in principle that special attention should be given to facilitate textile design protection. This impact on the 1999 Act is discussed in more detail in the Topics chapter under the heading "TRIPS".

There were several important details that had to be considered on the used of specimens. National offices might need to have a copy of the specimen while it was deferred, to consider in their novelty search, or for other reasons. A policy on how to handle the confidentiality of the deferred specimen had to be developed [Fifth Session Draft Act, note 8.06]. Another concern discussed was comparison of the substituted reproduction with the specimen to see that the designs were the same. The question of where this comparison should occur, either the International Bureau or the examining office, was discussed [Seventh Session Final Report #38].

In response to a question at the Diplomatic Conference, WIPO staff stated that specimens would be accessible after registration. They would not be disposed of and they would be kept in the file [WIPO Records of the Diplomatic Conference ... Hague Agreement (Geneva) 489 (#879-880) (2002)]. This step will be useful as a backup in case a comparison with the substitute reproduction is needed, and the national office does not have a copy or cannot locate the copy they had.

The use of specimens added significant advantages to the Draft Act for two-dimensional design owners and provided the necessary interfaces for all national offices. The comparison of the specimen and the subsequent replacement reproduction would be a matter for the examining office. Specimens could be filed in deferred multiple design applications giving considerable flexibility and providing lower initial registration costs. The Diplomatic Conference accepted this approach.

1999 Act Procedures for Two-Dimensional Design Protection

A number of topics are involved in the use of specimens, including use of multiple designs in an International Registration, the special treatment of specimen based applications in the fee structure, deferment in general, and use of reproductions. The present section will describe the major aspects of these relationships for the 1999 Act, and the Topics chapter on the specific subject should be consulted for more details.

A two-dimensional design can be protected under the 1999 Act in the same manner as any other design, with the filing of an application containing a reproduction and publication of the International Registration [1999 Act, art. 5(1)]. This routine procedure was not the only way to protect two-dimensional designs. [1999 Act, art. 5(1)(iii) and rule 16(3)]. The advantages of using a specimen and requesting deferment of the International Registration should be considered.

The initial filing of a specimen, instead of a reproduction, to obtain a filing date for a deferred International Registration set up a unique set of options that over time could be used as a very effective way to reduce the cost of two-dimensional design protection. In an industry with a lot of new designs that are untested in the marketplace, a procedure that allowed time to consider which designs would be useful before further expenditure of money for protection could be a very useful approach.

The 1999 Act provided several filing options, by allowing up to 30 months of deferment, in some contracting parties, to review which designs were most valuable and needed continued protection [1999 Act, rule 16(1)]. During the time that the International Registration was deferred, the costs for protecting many designs would be much lower with the specimens, saving the reproduction cost. The deferment allowed the publication fee to be deferred, so that only when the deferment period was close to the end did a decision have to be made on which designs to protect [1999 Act, rule 16(3)]. Then, a reproduction was substituted for the specimen and the publication fee was paid. Any initially protected design that was no longer valuable, or where protection in a specific country was not needed, could be renounced and the publication fee would not be paid [1999 Act, art. 11(5), and Administrative Instructions, sec. 601]. It was a very flexible set up for managing effectively the cost of a design protection program.

The advantage of the 1999 Act for two-dimensional designs was even more enhanced by the feature of allowing in one International Registration up to 100 designs in a class [1999 Act, art. 5(4) and rule 7(3)(iii)]. Since two-dimensional designs usually are generated in large numbers, especially in the textile and fashion industries, the multiple design feature increases the benefits of the basic specimen and deferment approach.

The need to have a copy of the deferred design specimen to use as part of the examination office search was the same for any deferred International Registration. A procedure was set up to ensure that a designated contracting party office could notify the International Bureau that it wanted to receive copies of all deferred International Registrations [1999 Act, art. 10(5)]. The applicant provided the extra copies to send at the time of filing the International Application [1999 Act, rule 10(1)(ii)]. The question of what effect the deferred International Registration had as prior art was a matter decided under national law.

Specimen Special Filing Procedures

The 1999 Act had special filing procedures for specimens, to keep the handling of them manageable [1999 Act, rule 10(2), Administrative Instructions sec. 406]. The specimens must be packaged in a certain way for receipt by the International Bureau. The maximum dimensions and weight of the package were specified [1999 Act, Administrative Instructions sec. 406]. Clearly this procedure was set up initially for the convenience of textile design owners that file fabric samples. Other industries desiring to protect two-dimensional designs should be able to work out their own arrangements, if needed.

Term for Protection

Meetings of Experts and Diplomatic Conference

The first Draft Act proposed a minimum 10-year protection term from the date of the International Registration, and the protection term maximum would correspond to the national design registration term [Second Session Draft Act, art. 10]. A majority of experts favored a minimum of 15 years from the International Registration date and the national maximum term [Second Session Final Report #101]. This recommendation was approved by the Diplomatic Conference [WIPO Records of the Diplomatic Conference ... Hague Agreement (Geneva) 455-458 (#538-566) (2002)]. The proposal would require some countries to increase their design registration protection term. The trend in 1991, at the time of the Second Session, was for longer design protection terms, and this is the current trend.

One concern at the experts meetings was to satisfy those national offices that were close to, but may be slightly less than 15 years protection, like the U.S. 14 years protection from grant of rights, after examination. In the U.S. compliance would depend on how fast the examination process took. U.S. design registration examination could be expedited and rights granted in less than a year from receipt of the application. In this case, the time from the International Registration date to the protection term end could be less than 15 years. The U.S. delegation and other experts recognized the need to review this timing requirement, and a change in U.S. law would be required, probably to a single term running from the U.S. filing date, and this type of change was indicated as acceptable [Fourth Session Final Report #129]. There was a willingness shown by the U.S. delegation to make the U.S. design patent law compatible with the 15-year minimum term. A similar change was made in U.S. Utility patent law, which based the protection term on the U.S. filing date [35 U.S.C. sec. 154(2) (2004)].

The discussion pointed out also that renunciation of International Registration effect was available, to shorten the term of protection, if an issue of double patenting required limiting the International Registration term. It would be important to have the International Bureau allow a statement shortening the term of an International Registration effect, to allow for circumstances that require this step. The experts meeting discussion suggested that this concern could be met [Fifth Session Final Report #106].

Other related topics discussed at the Diplomatic Conference were fee payment and renewal. The pros and cons of having a single designation fee for the full protection term were debated. It was stated that the Draft Act permitted that procedure, but users would like shorter term protection and renewal. In the Draft Act there was flexibility for the contracting party, a concept of the Hague Agreement System that minimized the need for a change in the national design laws to become a member [WIPO Records of the Diplomatic Conference ... Hague Agreement (Geneva) 456-458 (#540-#559) (2002)].

1999 Act Protection Term

A minimum protection term of 15 years from International Registration date was provided in the 1999 Act [1999 Act, arts. 10(2) and 17(3)]. The initial term was five years, renewable at five year intervals [1999 Act, art. 7(2)], subject to special arrangements. The International Registration maximum term for a contracting party is the term of the national design registration, with renewal [1999 Act, art. 17(3)(b)]. Each contracting party has to notify International Bureau of the maximum protection terms under national law [1999 Act, art. 17(3)c].

The section on International Registration renewal has useful related information.

TRIPS

Meetings of Experts and Diplomatic Conference

An important influence on the Meetings of Experts was the TRIPS statement concerning textile designs. It stated:

> "Each Member shall ensure that requirements for securing protection for textile designs, in particular in regard to any cost, examination or publication, do not unreasonable impair the opportunity to seek and obtain such protection. Members shall be free to meet this obligation through industrial design law or through copyright law". [TRIPS, art. 25(2)]

Professor Gervais, in his book on TRIPS explained the history of this provision, observing that "Art. 25(2) addresses the need of the textiles industry to protect large numbers of designs relatively quickly ..." [Daniel Gervais, The TRIPS Agreement Drafting History and Analysis 213 (2003)]. He also noted that there was no precise definition for the meaning of the word "textiles" from the history of the agreement, and specialized WTO agreements should be consulted for an acceptable international definition, Id. 214.

The textile industry representatives at the experts meetings often referred to this TRIPS policy statement as the guideline for the Hague Agreement revision [*see*, for example, Fifth Session Final Report #113-#116].

This support for textile design protection implementation was evidenced by the Meeting of Experts discussion favoring use of specimens as the initial filing for two-dimensional designs. The concept of a short term, low cost design protection system based on a deferred International Registration were other steps taken in the direction recommended by the TRIPS art. 25(2). Other types of designs benefitted from this effort to achieve lower cost textile design protection.

1999 Act Textile Design Protection

Several features in the 1999 Act provided effective protection for textile designs, consistent with the above-quoted TRIPS art. 25(2). These features will be discussed briefly here. More details can be found in the Topics chapter.

The 1999 Act has procedures that facilitate filing a multiple design International Registration application with a large number of designs at low cost. There is a 30-month maximum period of deferment, depending on national laws, to keep a design confidential. During the deferment there were protection rights from the International Registration, and infringement of these designs could be addressed. The procedural flexibility to select designs for continued protection, by completing the publication step only for valuable designs, gives the design owner a way to reduce overall protection costs.

Unity of Design

Meeting of Experts and Diplomatic Conference

In the Fourth Meeting of Experts the concept of unity of design was addressed [Fourth Session Draft Act, art. 4(4)(b), note 4.18]. The experts recognized that some countries, particularly ones with examining offices, allowed only one inventive design concept in each registration, following a practice found in patent law. It was argued by experts from the examining countries that a single design concept in an application allowed more efficient novelty examination and classification. This practice was in contrast with systems that allowed multiple designs in a registration that have no related design concepts. These systems differed significantly, and the Hague Agreement revision would not be able to change this practice at this time.

The experts worked to accommodate these systems, by recommending that a filing date would be obtained for filing a multiple design International Registration in a contracting party. The International Registration would be divided into separate parts for each design concept, reviewed independently and each part would retain the original filing date. The negative side of this compromise was that each new part would require a designation fee and other fees as a separate design. This practice was the same as for national application with multiple design concepts, so no unfairness was present. The Fifth Meeting of Experts completed, essentially, the work on interfacing the unity of design practice into the Draft Act and the Diplomatic Conference accepted this proposal [Fifth Session Draft Act, art. 4(b), notes 4.15-4.17; [WIPO Records of the Diplomatic Conference ... Hague Agreement (Geneva) 459-461 (#576-#593) (2002)].

1999 Act Interface with Unity Design Concept

The unity of design topic is closely related to the topics of multiple designs and fees. It is suggested that a review of these related topics in this chapter would be helpful.

A contracting party that has a design system using the unity of design principle has to inform the WIPO Director General through a declaration [1999 Act, art. 13(1)]. The 1999 Act states that an International Registration with multiple designs can be filed in a contracting party. Upon refusal, the national office would divide the application into separate design concept files for review, with the same filing date as the parent International Registration [1999 Act, arts. 12(1) and 13(3)]. This approved practice corresponded to how multiple design national applications were handled.

The division procedure for multiple design International Registration will result in new information on each separate part of the original International Registration, and this information must be communicated by the national office to the International Bureau [1999 Act, Administrative Instructions sec. 502]. The Bureau will publish this new information at the appropriate time.

APPENDIX I

Geneva Act of July 2, 1999

Table of Contents

INTRODUCTORY PROVISIONS

Article 1
Abbreviated Expressions

For the purposes of this Act:

(i) "the Hague Agreement" means the Hague Agreement Concerning the International Deposit of Industrial Designs, henceforth renamed the Hague Agreement Concerning the International Registration of Industrial Designs;

(ii) "this Act" means the Hague Agreement as established by the present Act;

(iii) "Regulations" means the Regulations under this Act;

(iv) "prescribed" means prescribed in the Regulations;

(v) "Paris Convention" means the Paris Convention for the Protection of Industrial Property, signed at Paris on March 20, 1883, as revised and amended;

(vi) "international registration" means the international registration of an industrial design effected according to this Act;

(vii) "international application" means an application for international registration;

(viii) "International Register" means the official collection of data concerning international registrations maintained by the International Bureau, which data this Act or the Regulations require or permit to be recorded, regardless of the medium in which such data are stored;

(ix) "person" means a natural person or a legal entity;

(x) "applicant" means the person in whose name an international application is filed;

(xi) "holder" means the person in whose name an international registration is recorded in the International Register;

(xii) "intergovernmental organization" means an intergovernmental organization eligible to become party to this Act in accordance with Article 27(1)(ii);

(xiii) "Contracting Party" means any State or intergovernmental organization party to this Act;

(xiv) "applicant's Contracting Party" means the Contracting Party or one of the Contracting Parties from which the applicant derives its entitlement to file an international application by virtue of satisfying, in relation to that Contracting Party, at least one of the conditions specified in Article 3; where there are two or more Contracting Parties from which the applicant may, under Article 3, derive its entitlement to file an international application, "applicant's Contracting Party" means the one which, among those Contracting Parties, is indicated as such in the international application;

(xv) "territory of a Contracting Party" means, where the Contracting Party is a State, the territory of that State and, where the Contracting Party is an intergovernmental organization, the territory in which the constituent treaty of that intergovernmental organization applies;

(xvi) "Office" means the agency entrusted by a Contracting Party with the grant of protection for industrial designs with effect in the territory of that Contracting Party;

(xvii) "Examining Office" means an Office which *ex officio* examines applications filed with it for the protection of industrial designs at least to determine whether the industrial designs satisfy the condition of novelty;

(xviii) "designation" means a request that an international registration have effect in a Contracting Party; it also means the recording, in the International Register, of that request;

(xix) "designated Contracting Party" and "designated Office" means the Contracting Party and the Office of the Contracting Party, respectively, to which a designation applies;

(xx) "1934 Act" means the Act signed at London on June 2, 1934, of the Hague Agreement;

(xxi) "1960 Act" means the Act signed at The Hague on November 28, 1960, of the Hague Agreement;

(xxii) "1961 Additional Act" means the Act signed at Monaco on November 18, 1961, additional to the 1934 Act;

(xxiii) "Complementary Act of 1967" means the Complementary Act signed at Stockholm on July 14, 1967, as amended, of the Hague Agreement;

(xxiv) "Union" means the Hague Union established by the Hague Agreement of November 6, 1925, and maintained by the 1934 and 1960 Acts, the 1961 Additional Act, the Complementary Act of 1967 and this Act;

(xxv) "Assembly" means the Assembly referred to in Article 21(1)(a) or any body replacing that Assembly;

(xxvi) "Organization" means the World Intellectual Property Organization;

(xxvii) "Director General" means the Director General of the Organization;

(xxviii) "International Bureau" means the International Bureau of the Organization;

(xxix) "instrument of ratification" shall be construed as including instruments of acceptance or approval.

Article 2
Applicability of Other Protection Accorded by Laws of Contracting Parties and by Certain International Treaties

(1) [*Laws of Contracting Parties and Certain International Treaties*] The provisions of this Act shall not affect the application of any greater protection which may be accorded by the law of a Contracting Party, nor shall they affect in any way the protection accorded to works of art and works of applied art by international copyright treaties and conventions, or the protection accorded to industrial designs under the Agreement on Trade-Related Aspects of Intellectual Property Rights annexed to the Agreement Establishing the World Trade Organization.

(2) [*Obligation to Comply with the Paris Convention*] Each Contracting Party shall comply with the provisions of the Paris Convention which concern industrial designs.

Chapter I
International Application and International Registration

Article 3
Entitlement to File an International Application

Any person that is a national of a State that is a Contracting Party or of a State member of an intergovernmental organization that is a Contracting Party, or that has a domicile, a habitual residence or a real and effective industrial or commercial establishment in the territory of a Contracting Party, shall be entitled to file an international application.

Article 4
Procedure for Filing the International Application

(1) [*Direct or Indirect Filing*]

(a) The international application may be filed, at the option of the applicant, either directly with the International Bureau or through the Office of the applicant's Contracting Party.

(b) Notwithstanding subparagraph (a), any Contracting Party may, in a declaration, notify the Director General that international applications may not be filed through its Office.

(2) [*Transmittal Fee in Case of Indirect Filing*] The Office of any Contracting Party may require that the applicant pay a transmittal fee to it, for its own benefit, in respect of any international application filed through it.

Article 5
Contents of the International Application

(1) [*Mandatory Contents of the International Application*] The international application shall be in the prescribed language or one of the prescribed languages and shall contain or be accompanied by

(i) a request for international registration under this Act;

(ii) the prescribed data concerning the applicant;

(iii) the prescribed number of copies of a reproduction or, at the choice of the applicant, of several different reproductions of the industrial design that is the subject of the international application, presented in the prescribed manner; however, where the industrial design is two-dimensional and a request for deferment of publication is made in accordance with paragraph (5), the international application may, instead of containing reproductions, be accompanied by the prescribed number of specimens of the industrial design;

(iv) an indication of the product or products which constitute the industrial design or in relation to which the industrial design is to be used, as prescribed;

(v) an indication of the designated Contracting Parties;

(vi) the prescribed fees;

(vii) any other prescribed particulars.

(2) [*Additional Mandatory Contents of the International Application*]

(a) Any Contracting Party whose Office is an Examining Office and whose law, at the time it becomes party to this Act, requires that an application for the grant of protection to an industrial design contain any of the elements specified in subparagraph (b) in order for that application to be accorded a filing date under that law may, in a declaration, notify the Director General of those elements.

(b) The elements that may be notified pursuant to subparagraph (a) are the following:

(i) indications concerning the identity of the creator of the industrial design that is the subject of that application;

(ii) a brief description of the reproduction or of the characteristic features of the industrial design that is the subject of that application;

(iii) a claim.

(c) Where the international application contains the designation of a Contracting Party that has made a notification under subparagraph (a), it shall also contain, in the prescribed manner, any element that was the subject of that notification.

(3) [*Other Possible Contents of the International Application*] The international application may contain or be accompanied by such other elements as are specified in the Regulations.

(4) [*Several Industrial Designs in the Same International Application*] Subject to such conditions as may be prescribed, an international application may include two or more industrial designs.

(5) [*Request for Deferred Publication*] The international application may contain a request for deferment of publication.

Article 6
Priority

(1) [*Claiming of Priority*]

(a) The international application may contain a declaration claiming, under Article 4 of the Paris Convention, the priority of one or more earlier applications filed in or for any country party to that Convention or any Member of the World Trade Organization.

(b) The Regulations may provide that the declaration referred to in subparagraph (a) may be made after the filing of the international application. In such case, the Regulations shall prescribe the latest time by which such declaration may be made.

(2) [*International Application Serving as a Basis for Claiming Priority*] The international application shall, as from its filing date and whatever may be its subsequent fate, be equivalent to a regular filing within the meaning of Article 4 of the Paris Convention.

Article 7
Designation Fees

(1) [*Prescribed Designation Fee*] The prescribed fees shall include, subject to paragraph (2), a designation fee for each designated Contracting Party.

(2) [*Individual Designation Fee*] Any Contracting Party whose Office is an Examining Office and any Contracting Party that is an intergovernmental organization may, in a declaration, notify the Director General that, in connection with any international application in which it is designated, and in connection with the renewal of any international registration resulting from such an international application, the prescribed designation fee referred to in paragraph (1) shall be replaced by an individual designation fee, whose amount shall be indicated in the declaration and can be changed in further declarations. The said amount may be fixed by the said Contracting Party for the initial term of protection and for each term of renewal or for the maximum period of protection allowed by the Contracting Party concerned. However, it may not be higher than the equivalent of the amount which the Office of that Contracting Party would be entitled to receive from an applicant for a grant of protection for an equivalent period to the same number of industrial designs, that amount being diminished by the savings resulting from the international procedure.

(3) [*Transfer of Designation Fees*] The designation fees referred to in paragraphs (1) and (2) shall be transferred by the International Bureau to the Contracting Parties in respect of which those fees were paid.

Article 8
Correction of Irregularities

(1) [*Examination of the International Application*] If the International Bureau finds that the international application does not, at the time of its receipt by the International Bureau, fulfill the requirements of this Act and the Regulations, it shall invite the applicant to make the required corrections within the prescribed time limit.

(2) [*Irregularities Not Corrected*]

(a) If the applicant does not comply with the invitation within the prescribed time limit, the international application shall, subject to subparagraph (b), be considered abandoned.

(b) In the case of an irregularity which relates to Article 5(2) or to a special requirement notified to the Director General by a Contracting Party in accordance with the Regulations, if the applicant does not comply with the invitation within the prescribed time limit, the international application shall be deemed not to contain the designation of that Contracting Party.

Article 9
Filing Date of the International Application

(1) [*International Application Filed Directly*] Where the international application is filed directly with the International Bureau, the filing date shall, subject to paragraph (3), be the date on which the International Bureau receives the international application.

(2) [*International Application Filed Indirectly*] Where the international application is filed through the Office of the applicant's Contracting Party, the filing date shall be determined as prescribed.

(3) [*International Application with Certain Irregularities*] Where the international application has, on the date on which it is received by the International Bureau, an irregularity which is prescribed as an irregularity entailing a postponement of the filing date of the international application, the filing date shall be the date on which the correction of such irregularity is received by the International Bureau.

Article 10[1]
International Registration, Date of the International Registration, Publication and Confidential Copies of the International Registration

(1) [*International Registration*] The International Bureau shall register each industrial design that is the subject of an international application immediately upon receipt by it of the international application or, where corrections are invited under Article 8, immediately upon receipt of the required corrections. The registration shall be effected whether or not publication is deferred under Article 11.

(2) [*Date of the International Registration*]

(a) Subject to subparagraph (b), the date of the international registration shall be the filing date of the international application.

(b) Where the international application has, on the date on which it is received by the International Bureau, an irregularity which relates to Article 5(2), the date of the international registration shall be the date on which the correction of such irregularity is received by the International Bureau or the filing date of the international application, whichever is the later.

(3) [*Publication*]

(a) The international registration shall be published by the International Bureau. Such publication shall be deemed in all Contracting Parties to be sufficient publicity, and no other publicity may be required of the holder.

(b) The International Bureau shall send a copy of the publication of the international registration to each designated Office.

(4) [*Maintenance of Confidentiality Before Publication*] Subject to paragraph (5) and Article 11(4)(b), the International Bureau shall keep in confidence each international application and each international registration until publication.

(5) [*Confidential Copies*]

(a) The International Bureau shall, immediately after registration has been effected, send a copy of the international registration, along with any relevant statement, document or specimen accompanying the international application, to each Office that has notified the International Bureau that it

---

1 When adopting Article 10, the Diplomatic Conference understood that nothing in this Article precludes access to the international application or the international registration by the applicant or the holder or a person having the consent of the applicant or the holder.

wishes to receive such a copy and has been designated in the international application.

(b) The Office shall, until publication of the international registration by the International Bureau, keep in confidence each international registration of which a copy has been sent to it by the International Bureau and may use the said copy only for the purpose of the examination of the international registration and of applications for the protection of industrial designs filed in or for the Contracting Party for which the Office is competent. In particular, it may not divulge the contents of any such international registration to any person outside the Office other than the holder of that international registration, except for the purposes of an administrative or legal proceeding involving a conflict over entitlement to file the international application on which the international registration is based. In the case of such an administrative or legal proceeding, the contents of the international registration may only be disclosed in confidence to the parties involved in the proceeding who shall be bound to respect the confidentiality of the disclosure.

Article 11
Deferment of Publication

(1) [Provisions of Laws of Contracting Parties Concerning Deferment of Publication]

(a) Where the law of a Contracting Party provides for the deferment of the publication of an industrial design for a period which is less than the prescribed period, that Contracting Party shall, in a declaration, notify the Director General of the allowable period of deferment.

(b) Where the law of a Contracting Party does not provide for the deferment of the publication of an industrial design, the Contracting Party shall, in a declaration, notify the Director General of that fact.

(2) [*Deferment of Publication*] Where the international application contains a request for deferment of publication, the publication shall take place,

(i) where none of the Contracting Parties designated in the international application has made a declaration under paragraph (1), at the expiry of the prescribed period or,

(ii) where any of the Contracting Parties designated in the international application has made a declaration under paragraph (1)(a), at the expiry of the period notified in such declaration or, where there is more than one such designated Contracting Party, at the expiry of the shortest period notified in their declarations.

(3) [*Treatment of Requests for Deferment Where Deferment Is Not Possible Under Applicable Law*] Where deferment of publication has been requested and any of the Contracting Parties designated in the international application has made a declaration under paragraph (1)(b) that deferment of publication is not possible under its law,

(i) subject to item (ii), the International Bureau shall notify the applicant accordingly; if, within the prescribed period, the applicant does not, by notice in writing to the International Bureau, withdraw the designation of the said Contracting Party, the International Bureau shall disregard the request for deferment of publication;

(ii) where, instead of containing reproductions of the industrial design, the international application was accompanied by specimens of the industrial design, the International Bureau shall disregard the designation of the said Contracting Party and shall notify the applicant accordingly.

(4) [*Request for Earlier Publication or for Special Access to the International Registration*]

(a) At any time during the period of deferment applicable under paragraph (2), the holder may request publication of any or all of the industrial designs that are the subject of the international registration, in which case the period of deferment in respect of such industrial design or designs shall be considered to have expired on the date of receipt of such request by the International Bureau.

(b) The holder may also, at any time during the period of deferment applicable under paragraph (2), request the International Bureau to provide a third party specified by the holder with an extract from, or to allow such a party access to, any or all of the industrial designs that are the subject of the international registration.

(5) [*Renunciation and Limitation*]

(a) If, at any time during the period of deferment applicable under paragraph (2), the holder renounces the international registration in respect of all the designated Contracting Parties, the industrial design or designs that are the subject of the international registration shall not be published.

(b) If, at any time during the period of deferment applicable under paragraph (2), the holder limits the international registration, in respect of all of the designated Contracting Parties, to one or some of the industrial designs that are the subject of the international registration, the other industrial design or designs that are the subject of the international registration shall not be published.

(6) [*Publication and Furnishing of Reproductions*]

(a) At the expiration of any period of deferment applicable under the provisions of this Article, the International Bureau shall, subject to the payment of the prescribed fees, publish the international registration. If such fees are not paid as prescribed, the international registration shall be canceled and publication shall not take place.

(b) Where the international application was accompanied by one or more specimens of the industrial design in accordance with Article 5(1)(iii), the holder shall submit the prescribed number of copies of a reproduction of each industrial design that is the subject of that application to the International Bureau within the prescribed time limit. To the extent that the holder does not do so, the international registration shall be canceled and publication shall not take place.

Article 12
Refusal

(1) [*Right to Refuse*] The Office of any designated Contracting Party may, where the conditions for the grant of protection under the law of that Contracting Party are not met in respect of any or all of the industrial designs that are the subject of an international registration, refuse the effects, in part or in whole, of the international registration in the territory of the said Contracting Party, provided that no Office may refuse the effects, in part or in whole, of any international registration on the ground that requirements relating to the form or contents of the international application that are provided for in this Act or the Regulations or are additional to, or different from, those requirements have not been satisfied under the law of the Contracting Party concerned.

(2) [*Notification of Refusal*]

(a) The refusal of the effects of an international registration shall be communicated by the Office to the International Bureau in a notification of refusal within the prescribed period.

(b) Any notification of refusal shall state all the grounds on which the refusal is based.

(3) [*Transmission of Notification of Refusal; Remedies*]

(a) The International Bureau shall, without delay, transmit a copy of the notification of refusal to the holder.

(b) The holder shall enjoy the same remedies as if any industrial design that is the subject of the international registration had been the subject of an application for the grant of protection under the law applicable to the Office that communicated the refusal. Such remedies shall at least consist of the possibility of a re-examination or a review of the refusal or an appeal against the refusal.

(4)[2] [*Withdrawal of Refusal*] Any refusal may be withdrawn, in part or in whole, at any time by the Office that communicated it.

Article 13
Special Requirements Concerning Unity of Design

(1) [*Notification of Special Requirements*] Any Contracting Party whose law, at the time it becomes party to this Act, requires that designs that are the subject of the same application conform to a requirement of unity of design, unity of production or unity of use, or belong to the same set or composition of items, or that only one independent and distinct design may be claimed in a single application, may, in a declaration, notify the Director General accordingly. However, no such declaration shall affect the right of an applicant to include two or more industrial designs in an international application in accordance with Article 5(4), even if the application designates the Contracting Party that has made the declaration.

(2) [*Effect of Declaration*] Any such declaration shall enable the Office of the Contracting Party that has made it to refuse the effects of the international registration pursuant to Article 12(1) pending compliance with the requirement notified by that Contracting Party.

---

2 When adopting Article 12(4), Article 14(2)(b) and Rule 18(4), the Diplomatic Conference understood that a withdrawal of refusal by an Office that has communicated a notification of refusal may take the form of a statement to the effect that the Office concerned has decided to accept the effects of the international registration in respect of the industrial designs, or some of the industrial designs, to which the notification of refusal related. It was also understood that an Office may, within the period allowed for communicating a notification of refusal, send a statement to the effect that it has decided to accept the effects of the international registration even where it has not communicated such a notification of refusal.

(3) [*Further Fees Payable on Division of Registration*] Where, following a notification of refusal in accordance with paragraph (2), an international registration is divided before the Office concerned in order to overcome a ground of refusal stated in the notification, that Office shall be entitled to charge a fee in respect of each additional international application that would have been necessary in order to avoid that ground of refusal.

Article 14
Effects of the International Registration

(1) [*Effect as Application Under Applicable Law*] The international registration shall, from the date of the international registration, have at least the same effect in each designated Contracting Party as a regularly-filed application for the grant of protection of the industrial design under the law of that Contracting Party.

(2) [*Effect as Grant of Protection Under Applicable Law*]

(a) In each designated Contracting Party the Office of which has not communicated a refusal in accordance with Article 12, the international registration shall have the same effect as a grant of protection for the industrial design under the law of that Contracting Party at the latest from the date of expiration of the period allowed for it to communicate a refusal or, where a Contracting Party has made a corresponding declaration under the Regulations, at the latest at the time specified in that declaration.

(b)[3] Where the Office of a designated Contracting Party has communicated a refusal and has subsequently withdrawn, in part or in whole, that refusal, the international registration shall, to the extent that the refusal is withdrawn, have the same effect in that Contracting Party as a grant of protection for the industrial design under the law of the said Contracting Party at the latest from the date on which the refusal was withdrawn.

(c) The effect given to the international registration under this paragraph shall apply to the industrial design or designs that are the subject of that registration as received from the International Bureau by the designated Office or, where applicable, as amended in the procedure before that Office.

(3) [*Declaration Concerning Effect of Designation of Applicant's Contracting Party*]

(a) Any Contracting Party whose Office is an Examining Office may, in a declaration, notify the Director General that, where it is the applicant's

---

[3] See footnote on page xxx.

Contracting Party, the designation of that Contracting Party in an international registration shall have no effect.

(b) Where a Contracting Party having made the declaration referred to in subparagraph (a) is indicated in an international application both as the applicant's Contracting Party and as a designated Contracting Party, the International Bureau shall disregard the designation of that Contracting Party.

Article 15
Invalidation

(1) [*Requirement of Opportunity of Defense*] Invalidation, by the competent authorities of a designated Contracting Party, of the effects, in part or in whole, in the territory of that Contracting Party, of the international registration may not be pronounced without the holder having, in good time, been afforded the opportunity of defending his rights.

(2) [*Notification of Invalidation*] The Office of the Contracting Party in whose territory the effects of the international registration have been invalidated shall, where it is aware of the invalidation, notify it to the International Bureau.

Article 16
Recording of Changes and Other Matters Concerning International Registrations

(1) [*Recording of Changes and Other Matters*] The International Bureau shall, as prescribed, record in the International Register

(i) any change in ownership of the international registration, in respect of any or all of the designated Contracting Parties and in respect of any or all of the industrial designs that are the subject of the international registration, provided that the new owner is entitled to file an international application under Article 3,

(ii) any change in the name or address of the holder,

(iii) the appointment of a representative of the applicant or holder and any other relevant fact concerning such representative,

(iv) any renunciation, by the holder, of the international registration, in respect of any or all of the designated Contracting Parties,

(v) any limitation, by the holder, of the international registration, in respect of any or all of the designated Contracting Parties, to one or some of the industrial designs that are the subject of the international registration,

(vi) any invalidation, by the competent authorities of a designated Contracting Party, of the effects, in the territory of that Contracting Party, of the international registration in respect of any or all of the industrial designs that are the subject of the international registration,

(vii) any other relevant fact, identified in the Regulations, concerning the rights in any or all of the industrial designs that are the subject of the international registration.

(2) [*Effect of Recording in International Register*] Any recording referred to in items (i), (ii), (iv), (v), (vi) and (vii) of paragraph (1) shall have the same effect as if it had been made in the Register of the Office of each of the Contracting Parties concerned, except that a Contracting Party may, in a declaration, notify the Director General that a recording referred to in item (i) of paragraph (1) shall not have that effect in that Contracting Party until the Office of that Contracting Party has received the statements or documents specified in that declaration.

(3) [*Fees*] Any recording made under paragraph (1) may be subject to the payment of a fee.

(4) [*Publication*] The International Bureau shall publish a notice concerning any recording made under paragraph (1). It shall send a copy of the publication of the notice to the Office of each of the Contracting Parties concerned.

Article 17
Initial Term and Renewal of the International Registration and Duration of Protection

(1) [*Initial Term of the International Registration*] The international registration shall be effected for an initial term of five years counted from the date of the international registration.

(2) [*Renewal of the International Registration*] The international registration may be renewed for additional terms of five years, in accordance with the prescribed procedure and subject to the payment of the prescribed fees.

(3) [*Duration of Protection in Designated Contracting Parties*]

(a) Provided that the international registration is renewed, and subject to subparagraph (b), the duration of protection shall, in each of the designated Contracting Parties, be 15 years counted from the date of the international registration.

(b) Where the law of a designated Contracting Party provides for a duration of protection of more than 15 years for an industrial design for which protection has been granted under that law, the duration of protection shall, provided that the international registration is renewed, be the same as that provided for by the law of that Contracting Party.

(c) Each Contracting Party shall, in a declaration, notify the Director General of the maximum duration of protection provided for by its law.

(4) [*Possibility of Limited Renewal*] The renewal of the international registration may be effected for any or all of the designated Contracting Parties and for any or all of the industrial designs that are the subject of the international registration.

(5) [*Recording and Publication of Renewal*] The International Bureau shall record renewals in the International Register and publish a notice to that effect. It shall send a copy of the publication of the notice to the Office of each of the Contracting Parties concerned.

Article 18
Information Concerning Published International Registrations

(1) [*Access to Information*] The International Bureau shall supply to any person applying therefor, upon the payment of the prescribed fee, extracts from the International Register, or information concerning the contents of the International Register, in respect of any published international registration.

(2) [*Exemption from Legalization*] Extracts from the International Register supplied by the International Bureau shall be exempt from any requirement of legalization in each Contracting Party.

Chapter II
Administrative Provisions

Article 19
Common Office of Several States

(1) [*Notification of Common Office*] If several States intending to become party to this Act have effected, or if several States party to this Act agree to effect, the unification of their domestic legislation on industrial designs, they may notify the Director General

(i) that a common Office shall be substituted for the national Office of each of them, and

(ii) that the whole of their respective territories to which the unified legislation applies shall be deemed to be a single Contracting Party for the purposes of the application of Articles 1, 3 to 18 and 31 of this Act.

(2) [*Time at Which Notification Is to Be Made*] The notification referred to in paragraph (1) shall be made,

(i) in the case of States intending to become party to this Act, at the time of the deposit of the instruments referred to in Article 27(2);

(ii) in the case of States party to this Act, at any time after the unification of their domestic legislation has been effected.

(3) [*Date of Entry into Effect of the Notification*] The notification referred to in paragraphs (1) and (2) shall take effect,

(i) in the case of States intending to become party to this Act, at the time such States become bound by this Act;

(ii) in the case of States party to this Act, three months after the date of the communication thereof by the Director General to the other Contracting Parties or at any later date indicated in the notification.

Article 20
Membership of the Hague Union

The Contracting Parties shall be members of the same Union as the States party to the 1934 Act or the 1960 Act.

Article 21
Assembly

(1) [*Composition*]

(a) The Contracting Parties shall be members of the same Assembly as the States bound by Article 2 of the Complementary Act of 1967.

(b) Each member of the Assembly shall be represented in the Assembly by one delegate, who may be assisted by alternate delegates, advisors and experts, and each delegate may represent only one Contracting Party.

(c) Members of the Union that are not members of the Assembly shall be admitted to the meetings of the Assembly as observers.

(2) [*Tasks*]

(a) The Assembly shall

(i) deal with all matters concerning the maintenance and development of the Union and the implementation of this Act;

(ii) exercise such rights and perform such tasks as are specifically conferred upon it or assigned to it under this Act or the Complementary Act of 1967;

(iii) give directions to the Director General concerning the preparations for conferences of revision and decide the convocation of any such conference;

(iv) amend the Regulations;

(v) review and approve the reports and activities of the Director General concerning the Union, and give the Director General all necessary instructions concerning matters within the competence of the Union;

(vi) determine the program and adopt the biennial budget of the Union, and approve its final accounts;

(vii) adopt the financial regulations of the Union;

(viii) establish such committees and working groups as it deems appropriate to achieve the objectives of the Union;

(ix) subject to paragraph (1)(c), determine which States, intergovernmental organizations and non-governmental organizations shall be admitted to its meetings as observers;

(x) take any other appropriate action to further the objectives of the Union and perform any other functions as are appropriate under this Act.

(b) With respect to matters which are also of interest to other Unions administered by the Organization, the Assembly shall make its decisions after having heard the advice of the Coordination Committee of the Organization.

(3) [*Quorum*]

(a) One-half of the members of the Assembly which are States and have the right to vote on a given matter shall constitute a quorum for the purposes of the vote on that matter.

(b) Notwithstanding the provisions of subparagraph (a), if, in any session, the number of the members of the Assembly which are States, have the right to vote on a given matter and are represented is less than one-half but equal to or more than one-third of the members of the Assembly which are States and have the right to vote on that matter, the Assembly may make decisions but, with the exception of decisions concerning its own procedure, all such decisions shall take effect only if the conditions set forth hereinafter are fulfilled. The International Bureau shall communicate the said decisions to the members of the Assembly which are States, have the right to vote on the said matter and were not represented and shall invite them to express in writing their vote or abstention within a period of three months from the date of the communication. If, at the expiration of this period, the number of such members having thus expressed their vote or abstention attains the number of the members which was lacking for attaining the quorum in the session itself, such decisions shall take effect provided that at the same time the required majority still obtains.

(4) [*Taking Decisions in the Assembly*]

(a) The Assembly shall endeavor to take its decisions by consensus.

(b) Where a decision cannot be arrived at by consensus, the matter at issue shall be decided by voting. In such a case,

(i) each Contracting Party that is a State shall have one vote and shall vote only in its own name, and

(ii) any Contracting Party that is an intergovernmental organization may vote, in place of its Member States, with a number of votes equal to the number of its Member States which are party to this Act, and no such intergovernmental organization shall participate in the vote if any one of its Member States exercises its right to vote, and *vice versa*.

(c) On matters concerning only States that are bound by Article 2 of the Complementary Act of 1967, Contracting Parties that are not bound by the said Article shall not have the right to vote, whereas, on matters concerning only Contracting Parties, only the latter shall have the right to vote.

(5) [*Majorities*]

(a) Subject to Articles 24(2) and 26(2), the decisions of the Assembly shall require two-thirds of the votes cast.

(b) Abstentions shall not be considered as votes.

(6) [*Sessions*]

(a) The Assembly shall meet once in every second calendar year in ordinary session upon convocation by the Director General and, in the absence of exceptional circumstances, during the same period and at the same place as the General Assembly of the Organization.

(b) The Assembly shall meet in extraordinary session upon convocation by the Director General, either at the request of one-fourth of the members of the Assembly or on the Director General's own initiative.

(c) The agenda of each session shall be prepared by the Director General.

(7) [*Rules of Procedure*] The Assembly shall adopt its own rules of procedure.

Article 22
International Bureau

(1) [*Administrative Tasks*]

(a) International registration and related duties, as well as all other administrative tasks concerning the Union, shall be performed by the International Bureau.

(b) In particular, the International Bureau shall prepare the meetings and provide the secretariat of the Assembly and of such committees of experts and working groups as may be established by the Assembly.

(2) [*Director General*] The Director General shall be the chief executive of the Union and shall represent the Union.

(3) [*Meetings Other than Sessions of the Assembly*] The Director General shall convene any committee and working group established by the Assembly and all other meetings dealing with matters of concern to the Union.

(4) [*Role of the International Bureau in the Assembly and Other Meetings*]

(a) The Director General and persons designated by the Director General shall participate, without the right to vote, in all meetings of the Assembly, the committees and working groups established by the Assembly, and any other meetings convened by the Director General under the aegis of the Union.

(b) The Director General or a staff member designated by the Director General shall be ex officio secretary of the Assembly, and of the committees, working groups and other meetings referred to in subparagraph (a).

(5) [*Conferences*]

(a) The International Bureau shall, in accordance with the directions of the Assembly, make the preparations for any revision conferences.

(b) The International Bureau may consult with intergovernmental organizations and international and national non-governmental organizations concerning the said preparations.

(c) The Director General and persons designated by the Director General shall take part, without the right to vote, in the discussions at revision conferences.

(6) [*Other Tasks*] The International Bureau shall carry out any other tasks assigned to it in relation to this Act.

Article 23
Finances

(1) [*Budget*]

(a) The Union shall have a budget.

(b) The budget of the Union shall include the income and expenses proper to the Union and its contribution to the budget of expenses common to the Unions administered by the Organization.

(c) Expenses not attributable exclusively to the Union but also to one or more other Unions administered by the Organization shall be considered to be expenses common to the Unions. The share of the Union in such common expenses shall be in proportion to the interest the Union has in them.

(2) [*Coordination with Budgets of Other Unions*] The budget of the Union shall be established with due regard to the requirements of coordination with the budgets of the other Unions administered by the Organization.

(3) [*Sources of Financing of the Budget*] The budget of the Union shall be financed from the following sources:

(i) fees relating to international registrations;

(ii) charges due for other services rendered by the International Bureau in relation to the Union;

(iii) sale of, or royalties on, the publications of the International Bureau concerning the Union;

(iv) gifts, bequests and subventions;

(v) rents, interests and other miscellaneous income.

(4) [*Fixing of Fees and Charges; Level of the Budget*]

(a) The amounts of the fees referred to in paragraph (3)(i) shall be fixed by the Assembly on the proposal of the Director General. Charges referred to in paragraph 3(ii) shall be established by the Director General and shall be provisionally applied subject to approval by the Assembly at its next session.

(b) The amounts of the fees referred to in paragraph (3)(i) shall be so fixed that the revenues of the Union from fees and other sources shall be at least sufficient to cover all the expenses of the International Bureau concerning the Union.

(c) If the budget is not adopted before the beginning of a new financial period, it shall be at the same level as the budget of the previous year, as provided in the financial regulations.

(5) [*Working Capital Fund*] The Union shall have a working capital fund which shall be constituted by the excess receipts and, if such excess does not suffice, by a single payment made by each member of the Union. If the fund becomes insufficient, the Assembly shall decide to increase it. The proportion and the terms of payment shall be fixed by the Assembly on the proposal of the Director General.

(6) [*Advances by Host State*]

(a) In the headquarters agreement concluded with the State on the territory of which the Organization has its headquarters, it shall be provided that, whenever the working capital fund is insufficient, such State shall grant advances. The amount of those advances and the conditions on which they are granted shall be the subject of separate agreements, in each case, between such State and the Organization.

(b) The State referred to in subparagraph (a) and the Organization shall each have the right to denounce the obligation to grant advances, by written notification. Denunciation shall take effect three years after the end of the year in which it has been notified.

(7) [*Auditing of Accounts*] The auditing of the accounts shall be effected by one or more of the States members of the Union or by external auditors, as provided in the financial regulations. They shall be designated, with their agreement, by the Assembly.

Article 24
Regulations

(1) [*Subject Matter*] The Regulations shall govern the details of the implementation of this Act. They shall, in particular, include provisions concerning

(i) matters which this Act expressly provides are to be prescribed;

(ii) further details concerning, or any details useful in the implementation of, the provisions of this Act;

(iii) any administrative requirements, matters or procedures.

(2) [*Amendment of Certain Provisions of the Regulations*]

(a) The Regulations may specify that certain provisions of the Regulations may be amended only by unanimity or only by a four-fifths majority.

(b) In order for the requirement of unanimity or a four-fifths majority no longer to apply in the future to the amendment of a provision of the Regulations, unanimity shall be required.

(c) In order for the requirement of unanimity or a four-fifths majority to apply in the future to the amendment of a provision of the Regulations, a four-fifths majority shall be required.

(3) [*Conflict Between This Act and the Regulations*] In the case of conflict between the provisions of this Act and those of the Regulations, the former shall prevail.

Chapter III
Revision and Amendment

Article 25
Revision of This Act

(1) [*Revision Conferences*] This Act may be revised by a conference of the Contracting Parties.

(2) [*Revision or Amendment of Certain Articles*] Articles 21, 22, 23 and 26 may be amended either by a revision conference or by the Assembly according to the provisions of Article 26.

Article 26
Amendment of Certain Articles by the Assembly

(1) [*Proposals for Amendment*]

(a) Proposals for the amendment by the Assembly of Articles 21, 22, 23 and this Article may be initiated by any Contracting Party or by the Director General.

(b) Such proposals shall be communicated by the Director General to the Contracting Parties at least six months in advance of their consideration by the Assembly.

(2) [*Majorities*] Adoption of any amendment to the Articles referred to in paragraph (1) shall require a three-fourths majority, except that adoption of any amendment to Article 21 or to the present paragraph shall require a four-fifths majority.

(3) [*Entry into Force*]

(a) Except where subparagraph (b) applies, any amendment to the Articles referred to in paragraph (1) shall enter into force one month after written notifications of acceptance, effected in accordance with their respective constitutional processes, have been received by the Director General from three-fourths of those Contracting Parties which, at the time the amendment was adopted, were members of the Assembly and had the right to vote on that amendment.

(b) Any amendment to Article 21(3) or (4) or to this subparagraph shall not enter into force if, within six months of its adoption by the Assembly, any Contracting Party notifies the Director General that it does not accept such amendment.

(c) Any amendment which enters into force in accordance with the provisions of this paragraph shall bind all the States and intergovernmental organizations which are Contracting Parties at the time the amendment enters into force, or which become Contracting Parties at a subsequent date.

Chapter IV
Final Provisions

Article 27
Becoming Party to This Act

(1) [*Eligibility*] Subject to paragraphs (2) and (3) and Article 28,

(i) any State member of the Organization may sign and become party to this Act;

(ii) any intergovernmental organization which maintains an Office in which protection of industrial designs may be obtained with effect in the territory in which the constituting treaty of the intergovernmental organization applies may sign and become party to this Act, provided that at least one of the member States of the intergovernmental organization is a member of the Organization and provided that such Office is not the subject of a notification under Article 19.

(2) [*Ratification or Accession*] Any State or intergovernmental organization referred to in paragraph (1) may deposit

(i) an instrument of ratification if it has signed this Act, or

(ii) an instrument of accession if it has not signed this Act.

(3) [*Effective Date of Deposit*]

(a) Subject to subparagraphs (b) to (d), the effective date of the deposit of an instrument of ratification or accession shall be the date on which that instrument is deposited.

(b) The effective date of the deposit of the instrument of ratification or accession of any State in respect of which protection of industrial designs may be obtained only through the Office maintained by an intergovernmental organization of which that State is a member shall be the date on which the instrument of that intergovernmental organization is deposited if that date is later than the date on which the instrument of the said State has been deposited.

(c) The effective date of the deposit of any instrument of ratification or accession containing or accompanied by the notification referred to in Article 19 shall be the date on which the last of the instruments of the States members of the group of States having made the said notification is deposited.

(d) Any instrument of ratification or accession of a State may contain or be accompanied by a declaration making it a condition to its being considered as deposited that the instrument of one other State or one intergovernmental organization, or the instruments of two other States, or the instruments of one other State and one intergovernmental organization, specified by name and eligible to become party to this Act, is or are also deposited. The instrument containing or accompanied by such a declaration shall be considered to have been deposited on the day on which the condition indicated in the declaration is fulfilled. However, when an instrument specified in the declaration itself contains, or is itself accompanied by, a declaration of the said kind, that instrument shall be considered as deposited on the day on which the condition specified in the latter declaration is fulfilled.

(e) Any declaration made under paragraph (d) may be withdrawn, in its entirety or in part, at any time. Any such withdrawal shall become effective on the date on which the notification of withdrawal is received by the Director General.

Article 28
Effective Date of Ratifications and Accessions

(1) [*Instruments to Be Taken into Consideration*] For the purposes of this Article, only instruments of ratification or accession that are deposited by States or intergovernmental organizations referred to in Article 27(1) and that have an effective date according to Article 27(3) shall be taken into consideration.

(2) [*Entry into Force of This Act*] This Act shall enter into force three months after six States have deposited their instruments of ratification or accession, provided that, according to the most recent annual statistics collected by the International Bureau, at least three of those States fulfill at least one of the following conditions:

(i) at least 3,000 applications for the protection of industrial designs have been filed in or for the State concerned, or

(ii) at least 1,000 applications for the protection of industrial designs have been filed in or for the State concerned by residents of States other than that State.

(3) [*Entry into Force of Ratifications and Accessions*]

(a) Any State or intergovernmental organization that has deposited its instrument of ratification or accession three months or more before the date of entry into force of this Act shall become bound by this Act on the date of entry into force of this Act.

(b) Any other State or intergovernmental organization shall become bound by this Act three months after the date on which it has deposited its instrument of ratification or accession or at any later date indicated in that instrument.

Article 29
Prohibition of Reservations

No reservations to this Act are permitted.

Article 30
Declarations Made by Contracting Parties

(1) [*Time at Which Declarations May Be Made*] Any declaration under Articles 4(1)(b), 5(2)(a), 7(2), 11(1), 13(1), 14(3), 16(2) or 17(3)(c) may be made

(i) at the time of the deposit of an instrument referred to in Article 27(2), in which case it shall become effective on the date on which the State or intergovernmental organization having made the declaration becomes bound by this Act, or

(ii) after the deposit of an instrument referred to in Article 27(2), in which case it shall become effective three months after the date of its receipt by the Director General or at any later date indicated in the declaration but shall apply only in respect of any international registration whose date of international registration is the same as, or is later than, the effective date of the declaration.

(2) [*Declarations by States Having a Common Office*] Notwithstanding paragraph (1), any declaration referred to in that paragraph that has been made by a State which has, with another State or other States, notified the Director General under Article 19(1) of the substitution of a common Office for their national Offices shall become effective only if that other State or those other States makes or make a corresponding declaration or corresponding declarations.

(3) [*Withdrawal of Declarations*] Any declaration referred to in paragraph (1) may be withdrawn at any time by notification addressed to the Director General. Such withdrawal shall take effect three months after the date on which the Director General has received the notification or at any later date indicated in the notification. In the case of a declaration made under Article 7(2), the withdrawal shall not affect international applications filed prior to the coming into effect of the said withdrawal.

Article 31
Applicability of the 1934 and 1960 Acts

(1) [*Relations Between States Party to Both This Act and the 1934 or 1960 Acts*] This Act alone shall be applicable as regards the mutual relations of States party to both this Act and the 1934 Act or the 1960 Act. However, such States shall, in their mutual relations, apply the 1934 Act or the 1960 Act, as the case may be, to industrial designs deposited at the International Bureau prior to the date on which this Act becomes applicable as regards their mutual relations.

(2) [*Relations Between States Party to Both This Act and the 1934 or 1960 Acts and States Party to the 1934 or 1960 Acts Without Being Party to This Act*]

(a) Any State that is party to both this Act and the 1934 Act shall continue to apply the 1934 Act in its relations with States that are party to the 1934 Act without being party to the 1960 Act or this Act.

(b) Any State that is party to both this Act and the 1960 Act shall continue to apply the 1960 Act in its relations with States that are party to the 1960 Act without being party to this Act.

Article 32
Denunciation of This Act

(1) [*Notification*] Any Contracting Party may denounce this Act by notification addressed to the Director General.

(2) [*Effective Date*] Denunciation shall take effect one year after the date on which the Director General has received the notification or at any later date indicated in the notification. It shall not affect the application of this Act to any international applica-

tion pending and any international registration in force in respect of the denouncing Contracting Party at the time of the coming into effect of the denunciation.

Article 33
Languages of This Act; Signature

(1) [*Original Texts; Official Texts*]

(a) This Act shall be signed in a single original in the English, Arabic, Chinese, French, Russian and Spanish languages, all texts being equally authentic.

(b) Official texts shall be established by the Director General, after consultation with the interested Governments, in such other languages as the Assembly may designate.

(2) [*Time Limit for Signature*] This Act shall remain open for signature at the headquarters of the Organization for one year after its adoption.

Article 34
Depositary

The Director General shall be the depositary of this Act.

APPENDIX II

Common Regulations Under the 1999 Act, the 1960 Act and the 1934 Act of the Hague Agreement (as in force on April 1, 2004)

Table of Contents

Chapter 1
General Provisions

Rule 1
Definitions

(1) [*Abbreviated Expressions*] For the purposes of these Regulations,

(i) "1999 Act" means the Act signed at Geneva on July 2, 1999, of the Hague Agreement;

(ii) "1960 Act" means the Act signed at The Hague on November 28, 1960, of the Hague Agreement;

(iii) "1934 Act" means the Act signed at London on June 2, 1934, of the Hague Agreement;

(iv) an expression which is used in these Regulations and is referred to in Article 1 of the 1999 Act has the same meaning as in that Act;

(v) "Administrative Instructions" means the Administrative Instructions referred to in Rule 34;

(vi) "communication" means any international application or any request, declaration, invitation, notification or information relating to or accompanying an international application or an international registration that is addressed to the Office of a Contracting Party, the International Bureau, the applicant or the holder by means permitted by these Regulations or the Administrative Instructions;

(vii) "official form" means a form established by the International Bureau or any form having the same contents and format;

(viii) "International Classification" means the Classification established under the Locarno Agreement Establishing an International Classification for Industrial Designs;

(ix) "prescribed fee" means the applicable fee set out in the Schedule of Fees;

(x) "Bulletin" means the periodical bulletin in which the International Bureau effects the publications provided for in the 1999 Act, the 1960 Act, the 1934 Act or these Regulations, whatever the medium used;

(xi) "Contracting Party designated under the 1999 Act" means a designated Contracting Party in respect of which the 1999 Act is applicable, either as the only common Act to which that designated Contracting Party and the applicant's Contracting Party are bound, or by virtue of Article 31(1), first sentence, of the 1999 Act;

(xii) "Contracting Party designated under the 1960 Act" means a designated Contracting Party in respect of which the 1960 Act is applicable, either as the only common Act to which that designated Contracting Party and the State of origin referred to in Article 2 of the 1960 Act are bound, or by virtue of Article 31(1), first sentence, of the 1960 Act;

(xiii) "Contracting Party designated under the 1934 Act" means a designated Contracting Party in respect of which the 1934 Act is applicable, as the only common Act to which that designated Contracting Party and the contracting country referred to in Article 1 of the 1934 Act are bound;

(xiv) "international application governed exclusively by the 1999 Act" means an international application in respect of which all designated Contracting Parties are Contracting Parties designated under the 1999 Act;

(xv) "international application governed exclusively by the 1960 Act" means an international application in respect of which all designated Contracting Parties are Contracting Parties designated under the 1960 Act;

(xvi) "international application governed exclusively by the 1934 Act" means an international application in respect of which all designated Contracting Parties are Contracting Parties designated under the 1934 Act;

(xvii) "international application governed by both the 1999 Act and the 1960 Act" means an international application in respect of which

- at least one Contracting Party has been designated under the 1999 Act,
- at least one Contracting Party has been designated under the 1960 Act, and
- no Contracting Party has been designated under the 1934 Act;

(xviii) "international application governed by both the 1999 Act and the 1934 Act" means an international application in respect of which

- at least one Contracting Party has been designated under the 1999 Act,
- at least one Contracting Party has been designated under the 1934 Act, and
- no Contracting Party has been designated under the 1960 Act;

(xix) "international application governed by both the 1960 Act and the 1934 Act" means an international application in respect of which

- at least one Contracting Party has been designated under the 1960 Act,
- at least one Contracting Party has been designated under the 1934 Act, and
- no Contracting Party has been designated under the 1999 Act;

(xx) "international application governed by the 1999 Act, the 1960 Act and the 1934 Act" means an international application in respect of which

- at least one Contracting Party has been designated under the 1999 Act,
- at least one Contracting Party has been designated under the 1960 Act, and

- at least one Contracting Party has been designated under the 1934 Act.

(2) [Correspondence Between Some Expressions Used in the 1999 Act, the 1960 Act and the 1934 Act] For the purposes of these Regulations,

(i) reference to "international application" or "international registration" shall be deemed, where appropriate, to include a reference to "international deposit" as referred to in the 1960 Act and the 1934 Act;

(ii) reference to "applicant" or "holder" shall be deemed, where appropriate, to include a reference to, respectively, "depositor" or "owner" as referred to in the 1960 Act and the 1934 Act;

(iii) reference to "Contracting Party" shall be deemed, where appropriate, to include a reference to a State party to the 1960 Act or to a country party to the 1934 Act;

(iv) reference to "Contracting Party whose Office is an examining Office" shall be deemed, where appropriate, to include a reference to "State having a novelty examination" as defined in Article 2 of the 1960 Act;

(v) reference to "individual designation fee" shall be deemed, where appropriate, to include a reference to the fee mentioned in Article 15(1)2(b) of the 1960 Act;

(vi) reference to "renewal" shall be deemed, where appropriate, to include a reference to "prolongation" referred to in the 1934 Act.

Rule 2
Communication with the International Bureau

Communications addressed to the International Bureau shall be effected as specified in the Administrative Instructions.

Rule 3
Representation Before the International Bureau

(1) [*Representative; Number Representatives*]

(a) The applicant or the holder may have a representative before the International Bureau.

(b) Only one representative may be appointed in respect of a given international application or international registration. Where the appointment indicates several representatives, only the one indicated first shall be considered to be a representative and be recorded as such.

(c) Where a partnership or firm composed of attorneys or patent or trademark agents has been indicated as representative to the International Bureau, it shall be regarded as one representative.

(2) [*Appointment of the Representative*]

(a) The appointment of a representative may be made in the international application, provided that the application is signed by the applicant.

(b) The appointment of a representative may also be made in a separate communication which may relate to one or more specified international applications or international registrations of the same applicant or holder. The said communication shall be signed by the applicant or the holder.

(c) Where the International Bureau considers that the appointment of a representative is irregular, it shall notify accordingly the applicant or holder and the purported representative.

(3) [*Recording and Notification of Appointment of a Representative; Effective Date of Appointment*]

(a) Where the International Bureau finds that the appointment of a representative complies with the applicable requirements, it shall record the fact that the applicant or holder has a representative, as well as the name and address of the representative, in the International Register. In such a case, the effective date of the appointment shall be the date on which the International Bureau received the international application or separate communication in which the representative is appointed.

(b) The International Bureau shall notify the recording referred to in subparagraph (a) to both the applicant or holder and the representative.

(4) [*Effect of Appointment of a Representative*]

(a) Except where these Regulations expressly provide otherwise, the signature of a representative recorded under paragraph (3)(a) shall replace the signature of the applicant or holder.

(b) Except where these Regulations expressly require that a communication be addressed to both the applicant or holder and the representative, the International Bureau shall address to the representative recorded under paragraph (3)(a) any communication which, in the absence of a representative, would have to be sent to the applicant or holder; any communication so addressed to the said representative shall have the same effect as if it had been addressed to the applicant or holder.

(c) Any communication addressed to the International Bureau by the representative recorded under paragraph (3)(a) shall have the same effect as if it had been addressed to the said Bureau by the applicant or holder.

(5) [*Cancellation of Recording; Effective Date of Cancellation*]

(a) Any recording under paragraph (3)(a) shall be canceled where cancellation is requested in a communication signed by the applicant, holder or representative. The recording shall be canceled ex officio by the International Bureau where a new representative is appointed or where a change in ownership is recorded and no representative is appointed by the new holder of the international registration.

(b) The cancellation shall be effective from the date on which the International Bureau receives the corresponding communication.

(c) The International Bureau shall notify the cancellation and its effective date to the representative whose recording has been canceled and to the applicant or holder.

Rule 4
Calculation of Time Limits

(1) [*Periods Expressed in Years*] Any period expressed in years shall expire, in the relevant subsequent year, in the month having the same name and on the day having the same number as the month and the day of the event from which the period starts to run, except that, where the event occurred on February 29 and in the relevant subsequent year February ends on the 28th, the period shall expire on February 28.

(2) [*Periods Expressed in Months*] Any period expressed in months shall expire, in the relevant subsequent month, on the day which has the same number as the day of the event from which the period starts to run, except that, where the relevant subsequent month has no day with the same number, the period shall expire on the last day of that month.

(3) [*Periods Expressed in Days*] The calculation of any period expressed in days shall start with the day following the day on which the relevant event occurred and shall expire accordingly.

(4) [*Expiry on a Day on Which the International Bureau or an Office Is Not Open to the Public*] If a period expires on a day on which the International Bureau or the Office concerned is not open to the public, the period shall, notwithstanding paragraphs (1) to (3), expire on the first subsequent day on which the International Bureau or the Office concerned is open to the public.

Rule 5
Irregularities in Postal and Delivery Services

(1) [*Communications Sent Through a Postal Service*] Failure by an interested party to meet a time limit for a communication addressed to the International Bureau and mailed through a postal service shall be excused if the interested party submits evidence showing, to the satisfaction of the International Bureau,

(i) that the communication was mailed at least five days prior to the expiry of the time limit, or, where the postal service was, on any of the ten days preceding the day of expiry of the time limit, interrupted on account of war, revolution, civil disorder, strike, natural calamity, or other like reason, that the communication was mailed not later than five days after postal service was resumed,

(ii) that the mailing of the communication was registered, or details of the mailing were recorded, by the postal service at the time of mailing, and

(iii) in cases where not all classes of mail normally reach the International Bureau within two days of mailing, that the communication was mailed by a class of mail which normally reaches the International Bureau within two days of mailing or by airmail.

(2) [*Communications Sent Through a Delivery Service*] Failure by an interested party to meet a time limit for a communication addressed to the International Bureau and sent through a delivery service shall be excused if the interested party submits evidence showing, to the satisfaction of the International Bureau,

(i) that the communication was sent at least five days prior to the expiry of the time limit, or, where the delivery service was, on any of the ten days preceding the day of expiry of the time limit, interrupted on account of war, revolution, civil disorder, natural calamity, or other like reason, that the communication was sent not later than five days after the delivery service was resumed, and

(ii) that details of the sending of the communication were recorded by the delivery service at the time of sending.

(3) [*Limitation on Excuse*] Failure to meet a time limit shall be excused under this Rule only if the evidence referred to in paragraph (1) or (2) and the communication or a duplicate thereof are received by the International Bureau not later than six months after the expiry of the time limit.

Rule 6
Languages

(1) [*International Application*] The international application shall be in English or French.

(2) [*Recording and Publication*] The recording in the International Register and the publication in the Bulletin of the international registration and of any data to be both recorded and published under these Regulations in respect of that international registration shall be in English and French. The recording and publication of the international registration shall indicate the language in which the international application was received by the International Bureau.

(3) [*Communications*] Any communication concerning an international application or the international registration resulting therefrom shall be

(i) in English or French where such communication is addressed to the International Bureau by the applicant or holder or by an Office;

(ii) in the language of the international application where the communication is addressed by the International Bureau to an Office, unless that Office has notified the International Bureau that all such communications are to be in English or that all such communications are to be in French;

(iii) in the language of the international application where the communication is addressed by the International Bureau to the applicant or holder unless the applicant or holder expresses the wish to receive all such communications in English although the international application was in French, or vice versa.

(4) [*Translation*] The translations needed for the recordings and publications under paragraph (2) shall be made by the International Bureau. The applicant may annex to the international application a proposed translation of any text matter contained in the international application. If the proposed translation is not considered by the International Bureau to be correct, it shall be corrected by the International Bureau after having invited the applicant to make, within one month from the invitation, observations on the proposed corrections.

Chapter 2
International Applications and International Registrations

Rule 7
Requirements Concerning the International Application

(1) [*Form and Signature*] The international application shall be presented on the official form. The international application shall be signed by the applicant.

(2) [*Fees*] The prescribed fees applicable to the international application shall be paid as provided for in Rules 27 and 28.

(3) [*Mandatory Contents of the International Application*] The international application shall contain or indicate

(i) the name of the applicant, given in accordance with the Administrative Instructions;

(ii) the address of the applicant, given in accordance with the Administrative Instructions;

(iii) the Contracting Party or Parties in respect of which the applicant fulfills the conditions to be the holder of an international registration;

(iv) the product or products which constitute the industrial design or in relation to which the industrial design is to be used, with an indication whether the product or products constitute the industrial design or are products in relation to which the industrial design is to be used; the product or products shall preferably be identified by using terms appearing in the list of goods of the International Classification;

(v) the number of industrial designs included in the international application, which may not exceed 100, and the number of reproductions or specimens of the industrial designs accompanying the international application in accordance with Rule 9 or 10;

(vi) the designated Contracting Parties;

(vii) the amount of the fees being paid and the method of payment, or instructions to debit the required amount of fees to an account opened with the International Bureau, and the identification of the party effecting the payment or giving the instructions.

(4) [*Additional Mandatory Contents of an International Application*]

(a) With respect to Contracting Parties designated under the 1999 Act in an international application, that application shall contain, in addition to the indications referred to in paragraph (3)(iii), the indication of the applicant's Contracting Party.

(b) Where a Contracting Party designated under the 1999 Act has notified the Director General, in accordance with Article 5(2)(a) of the 1999 Act, that its law requires one or more of the elements referred to in Article 5(2)(b) of the 1999 Act, the international application shall contain such element or elements, as prescribed in Rule 11.

(c) Where Rule 8 applies, the international application shall contain the indications referred to in Rule 8(2) and, where applicable, be accompanied by the statement or document referred to in that Rule.

(5) [*Optional Contents of an International Application*]

(a) An element referred to in item (i) or (ii) of Article 5(2)(b) of the 1999 Act or in Article 8(4)(a) of the 1960 Act may, at the option of the applicant, be included in the international application even where that element is not required in consequence of a notification in accordance with Article 5(2)(a) of the 1999 Act or in consequence of a requirement under Article 8(4)(a) of the 1960 Act.

(b) Where the applicant has a representative, the international application shall state the name and address of the representative, given in accordance with the Administrative Instructions.

(c) Where the applicant wishes, under Article 4 of the Paris Convention, to take advantage of the priority of an earlier filing, the international application shall contain a declaration claiming the priority of that earlier filing, together with an indication of the name of the Office where such filing was made and of the date and, where available, the number of that filing and, where the priority claim relates to less than all the industrial designs contained in the international application, the indication of those industrial designs to which the priority claim relates or does not relate.

(d) Where the applicant wishes to take advantage of Article 11 of the Paris Convention, the international application shall contain a declaration that the product or products which constitute the industrial design or in which the industrial design is incorporated have been shown at an official or officially recognized international exhibition, together with the place where the exhibition was held and the date on which the product or products were first exhibited there and, where less than all the industrial designs contained in the international application are concerned, the indication of those industrial designs to which the declaration relates or does not relate.

(e) Where the applicant wishes that publication of the industrial design be deferred, the international application shall contain a request for deferment of publication.

(f) The international application may also contain any declaration, statement or other relevant indication as may be specified in the Administrative Instructions.

(g) The international application may be accompanied by a statement that identifies information known by the applicant to be material to the eligibility for protection of the industrial design concerned.

(6) [*No Additional Matter*] If the international application contains any matter other than that required or permitted by the 1999 Act, the 1960 Act, the 1934 Act, these Regulations or the Administrative Instructions, the International Bureau shall delete it *ex officio*. If the international application is accompanied by any document other than those required or permitted, the International Bureau may dispose of the said document.

(7) [*All Products to Be in Same Class*] All the products which constitute the industrial designs to which an international application relates, or in relation to which the industrial designs are to be used, shall belong to the same class of the International Classification.

Rule 8
Special Requirements Concerning the Applicant

(1) [*Notification of Special Requirements*]

(a) Where the law of a Contracting Party bound by the 1999 Act requires that an application for the protection of an industrial design be filed in the name of the creator of the industrial design, that Contracting Party may, in a declaration, notify the Director General of that fact.

(b) The declaration referred to in subparagraph (a) shall specify the form and mandatory contents of any statement or document required for the purposes of paragraph (2).

(2) [*Identity of the Creator and Assignment of International Application*] Where an international application contains the designation of a Contracting Party that has made the declaration referred to in paragraph (1),

(i) it shall also contain indications concerning the identity of the creator of the industrial design, together with a statement, complying with the requirements specified in accordance with paragraph (1)(b), that the latter believes himself to be the creator of the industrial design; the person so identified as the creator shall be deemed to be the applicant for the purposes of the designation of that Contracting Party, irrespective of the person named as the applicant in accordance with Rule 7(3)(i);

(ii) where the person identified as the creator is a person other than the person named as the applicant in accordance with Rule 7(3)(i), the international application shall be accompanied by a statement or document, complying with the requirements specified in accordance with paragraph (1)(b), to the effect that it has been assigned by the person identified as the creator to the person named as the applicant. The latter person shall be recorded as the holder of the international registration.

Rule 9
Reproductions of the Industrial Design

(1) [*Form and Number of Reproductions of the Industrial Design*]

(a) Reproductions of the industrial design shall, at the option of the applicant, be in the form of photographs or other graphic representations of the industrial design itself or of the product or products which constitute the industrial design. The same product may be shown from different angles; views from different angles shall be included in different photographs or other graphic representations.

(b) Any reproduction shall be submitted in the number of copies specified in the Administrative Instructions.

(2) [*Requirements Concerning Reproductions*]

(a) Reproductions shall be of a quality permitting all the details of the industrial design to be clearly distinguished and permitting publication.

(b) Matter which is shown in a reproduction but for which protection is not sought may be indicated as provided for in the Administrative Instructions.

(3) [*Views Required*]

(a) Subject to subparagraph (b), any Contracting Party bound by the 1999 Act which requires certain specified views of the product or products which constitute the industrial design or in relation to which the industrial design is to be used shall, in a declaration, so notify the Director General, specifying the views that are required and the circumstances in which they are required.

(b) No Contracting Party may require more than one view where the industrial design or product is two-dimensional, or more than six views where the product is three-dimensional.

(4) [*Refusal on Grounds Relating to the Reproductions of the Industrial Design*] A Contracting Party may not refuse the effects of the international registration on the ground that requirements relating to the form of the reproductions of the industrial design that are additional to, or different from, those notified by that Contracting Party in accordance with paragraph (3)(a) have not been satisfied under its law. A Contracting Party may however refuse the effects of the international registration on the ground that the reproductions contained in the international registration are not sufficient to disclose fully the industrial design.

Rule 10
Specimens of the Industrial Design Where

Deferment of Publication Is Requested

(1) [*Number of Specimens*] Where an international application governed exclusively by the 1999 Act contains a request for deferment of publication in respect of a two-dimensional industrial design and, instead of being accompanied by the reproductions referred to in Rule 9, is accompanied by specimens of the industrial design, the following number of specimens shall accompany the international application:

(i) one specimen for the International Bureau, and

(ii) one specimen for each designated Office that has notified the International Bureau under Article 10(5) of the 1999 Act that it wishes to receive copies of international registrations.

(2) [*Specimens*] All the specimens shall be contained in a single package. The specimens may be folded. The maximum dimensions and weight of the package shall be specified in the Administrative Instructions.

Rule 11
Identity of Creator; Description; Claim

(1) [*Identity of Creator*] Where the international application contains indications concerning the identity of the creator of the industrial design, his name and address shall be given in accordance with the Administrative Instructions.

(2) [*Description*] Where the international application contains a description, the latter shall concern those features that appear in the reproductions of the industrial design and may not concern technical features of the operation of the industrial

design or its possible utilization. If the description exceeds 100 words, an additional fee, as set out in the Schedule of Fees, shall be payable.

(3) [*Claim*] A declaration under Article 5(2)(a) of the 1999 Act that the law of a Contracting Party requires a claim in order for an application for the grant of protection to an industrial design to be accorded a filing date under that law shall specify the exact wording of the required claim. Where the international application contains a claim, the wording of that claim shall be as specified in the said declaration.

Rule 12
Fees Concerning the International Application

(1) [*Prescribed Fees*]

(a) The international application shall be subject to the payment of the following fees:

(i) a basic fee;
(ii) a standard designation fee in respect of each designated Contracting Party that has not made a declaration under Article 7(2) of the 1999 Act or under Rule 36(1);
(iii) an individual designation fee in respect of each designated Contracting Party that has made a declaration under Article 7(2) of the 1999 Act or under Rule 36(1);
(iv) a publication fee.

(b) The amounts of the fees referred to in items (i), (ii) and (iv) are set out in the Schedule of Fees.

(2) [*When Fees to Be Paid*] The fees referred to in paragraph (1) are, subject to paragraph (3), payable at the time of filing the international application, except that, where the international application contains a request for deferment of publication, the publication fee may be paid later, in accordance with Rule 16(3)(a).

(3) [*Individual Designation Fee Payable in Two Parts*]

(a) A declaration under Article 7(2) of the 1999 Act or under Rule 36(1) may also specify that the individual designation fee to be paid in respect of the Contracting Party concerned comprises two parts, the first part to be paid at the time of filing the international application and the second part to be paid at a later date which is determined in accordance with the law of the Contracting Party concerned.

(b) Where subparagraph (a) applies, the reference in paragraph (1)(iii) to an individual designation fee shall be construed as a reference to the first part of the individual designation fee.

(c) The second part of the individual designation fee may be paid either directly to the Office concerned or through the International Bureau, at the option of the holder. Where it is paid directly to the Office concerned, the Office shall notify the International Bureau accordingly and the International Bureau shall record any such notification in the International Register. Where it is paid through the International Bureau, the International Bureau shall record the payment in the International Register and notify the Office concerned accordingly.

(d) Where the second part of the individual designation fee is not paid within the applicable period, the Office concerned shall notify the International Bureau and request the International Bureau to cancel the international registration in the International Register with respect to the Contracting Party concerned. The International Bureau shall proceed accordingly and so notify the holder.

Rule 13
International Application Filed Through an Office

(1) [*Date of Receipt by Office and Transmittal to the International Bureau*] Where an international application governed exclusively by the 1999 Act is filed through the Office of the applicant's Contracting Party, that Office shall notify the applicant of the date on which it received the application. At the same time as it transmits the international application to the International Bureau, the Office shall notify the International Bureau of the date on which it received the application. The Office shall notify the applicant of the fact that it has transmitted the international application to the International Bureau.

(2) [*Transmittal Fee*] An Office that requires a transmittal fee, as provided for in Article 4(2) of the 1999 Act, shall notify the International Bureau of the amount of such fee, which should not exceed the administrative costs of receiving and transmitting the international application, and its due date.

(3) [*Filing Date of International Application Filed Indirectly*] Subject to Rule 14(2), the filing date of an international application filed through an Office shall be

(i) where the international application is governed exclusively by the 1999 Act, the date on which the international application was received by that Office, provided that it is received by the International Bureau within one month of that date;

(ii) in any other case, the date on which the International Bureau receives the international application.

(4) [*Filing Date Where Applicant's Contracting Party Requires a Security Clearance*] Notwithstanding paragraph (3), a Contracting Party whose law, at the time that it becomes party to the 1999 Act, requires security clearance may, in a declaration, notify the Director General that the period of one month referred to in that paragraph shall be replaced by a period of six months.

Rule 14
Examination by the International Bureau

(1) [*Time Limit for Correcting Irregularities*] If the International Bureau finds that the international application does not, at the time of its receipt by the International Bureau, fulfill the applicable requirements, it shall invite the applicant to make the required corrections within three months from the date of the invitation sent by the International Bureau.

(2) [*Irregularities Entailing a Postponement of the Filing Date of the International Application*] Where the international application has, on the date on which it is received by the International Bureau, an irregularity which is prescribed as an irregularity entailing a postponement of the filing date of the international application, the filing date shall be the date on which the correction of such irregularity is received by the International Bureau. The irregularities which are prescribed as entailing a postponement of the filing date of the international application are the following:

(a) the international application is not in the prescribed language or one of the prescribed languages;

(b) any of the following elements is missing from the international application:

(i) an express or implicit indication that international registration under the 1999 Act, the 1960 Act or the 1934 Act is sought;
(ii) indications allowing the identity of the applicant to be established;
(iii) indications sufficient to enable the applicant or its representative, if any, to be contacted;
(iv) a reproduction, or, in accordance with Article 5(1)(iii) of the 1999 Act, a specimen, of each industrial design that is the subject of the international application;
(v) the designation of at least one Contracting Party.

(3) [*International Application Considered Abandoned; Reimbursement of Fees*] Where an irregularity, other than an irregularity referred to in Article 8(2)(b) of the 1999 Act, is not remedied within the time limit referred to in paragraph (1), the international application shall be considered abandoned and the International

Bureau shall refund any fees paid in respect of that application, after deduction of an amount corresponding to the basic fee.

Rule 15
Registration of the Industrial Design in the International Register

(1) [*Registration of the Industrial Design in the International Register*] Where the International Bureau finds that the international application conforms to the applicable requirements, it shall register the industrial design in the International Register and send a certificate to the holder.

(2) [*Contents of the Registration*] The international registration shall contain

(i) all the data contained in the international application, except any priority claim under Rule 7(5)(c) where the date of the earlier filing is more than six months before the filing date of the international application;

(ii) any reproduction of the industrial design;

(iii) the date of the international registration;

(iv) the number of the international registration;

(v) the relevant class of the International Classification, as determined by the International Bureau.

Rule 16
Deferment of Publication

(1) [*Maximum Period of Deferment*]

(a) The prescribed period for deferment of publication in respect of an international application governed exclusively by the 1999 Act shall be 30 months from the filing date or, where priority is claimed, from the priority date of the application concerned.

(b) The maximum period for deferment of publication in respect of an international application governed exclusively by the 1960 Act or by both the 1999 Act and the 1960 Act shall be 12 months from the filing date or, where priority is claimed, from the priority date of the application concerned.

(2) [*Period for Withdrawal of Designation Where Deferment Is Not Possible Under Applicable Law*] The period referred to in Article 11(3)(i) of the 1999 Act for the applicant to withdraw the designation of a Contracting Party whose law does not allow the deferment of publication shall be one month from the date of the notification sent by the International Bureau.

(3) [*Period for Paying Publication Fee and Submitting Reproductions*]

(a) The publication fee referred to in Rule 12(1)(a)(iv) shall be paid, and, where specimens have been submitted instead of reproductions in accordance with Rule 10, those reproductions shall be submitted, not later than three months before the period of deferment applicable under Article 11(2) of the 1999 Act or under Article 6(4)(a) of the 1960 Act expires or not later than three months before the period of deferment is considered to have expired in accordance with Article 11(4)(a) of the 1999 Act or with Article 6(4)(b) of the 1960 Act.

(b) Six months before the expiry of the period of deferment of publication referred to in subparagraph (a), the International Bureau shall, by sending an unofficial notice, remind the holder of the international registration, where applicable, of the date by which the publication fee referred to in paragraph (3) shall be paid and the reproductions referred to in paragraph (3) shall be submitted.

(4) [*Registration of Reproductions*] The International Bureau shall record in the International Register any reproduction submitted under paragraph (3) .

(5) [*Requirements Not Complied With*] If the requirements of paragraph (3) are not complied with, the international registration shall be canceled and shall not be published.

Rule 17
Publication of the International Registration

(1) [*Timing of Publication*] The international registration shall be published

(i) where the applicant so requests, immediately after the registration,

(ii) where deferment of publication has been requested and the request has not been disregarded, immediately after the date on which the period of deferment expired or is considered to have expired,

(iii) in any other case, six months after the date of the international registration or as soon as possible thereafter.

(2) [*Contents of Publication*] The publication of the international registration in the Bulletin shall contain

(i) the data recorded in the International Register;

(ii) the reproduction or reproductions of the industrial design;

(iii) where publication has been deferred, an indication of the date on which the period of deferment expired or is considered to have expired.

Chapter 3
Refusals and Invalidations

Rule 18
Notification of Refusal

(1) [*Period for Notification of Refusal*]

(a) The prescribed period for the notification of refusal of the effects of an international registration in accordance with Article 12(2) of the 1999 Act or Article 8(1) of the 1960 Act shall be six months from the publication of the international registration as provided for by Rule 26(3).

(b) Notwithstanding subparagraph (a), any Contracting Party whose Office is an Examining Office, or whose law provides for the possibility of opposition to the grant of protection, may, in a declaration, notify the Director General that, where it is designated under the 1999 Act, the period of six months referred to in that subparagraph shall be replaced by a period of 12 months.

(c) The declaration referred to in subparagraph (b) may also state that the international registration shall produce the effect referred to in Article 14(2)(a) of the 1999 Act at the latest

(i) at a time specified in the declaration which may be later than the date referred to in that Article but which shall not be more than six months after the said date or

(ii) at a time at which protection is granted according to the law of the Contracting Party where a decision regarding the grant of protection was unintentionally not communicated within the period applicable under subparagraph (a) or (b); in such a case, the Office of the Contracting Party concerned shall notify the International Bureau accordingly and endeavor to communicate such decision to the holder of the international registration concerned promptly thereafter.

(2) [*Notification of Refusal*]

(a) The notification of any refusal shall relate to one international registration, shall be dated and shall be signed by the Office making the notification.

(b) The notification shall contain or indicate

(i) the Office making the notification,

(ii) the number of the international registration,

(iii) all the grounds on which the refusal is based together with a reference to the corresponding essential provisions of the law,

(iv) where the grounds on which the refusal is based refer to similarity with an industrial design which has been the subject of an earlier national, regional or international application or registration, the filing date and number, the priority date (if any), the registration date and number (if available), a copy of a reproduction of the earlier industrial design (if that reproduction is accessible to the public) and the name and address of the owner of the said industrial design, as provided for in the Administrative Instructions,

(v) where the refusal does not relate to all the industrial designs that are the subject of the international registration, those to which it relates or does not relate,

(vi) whether the refusal may be subject to review or appeal and, if so, the time limit, reasonable under the circumstances, for any request for review of, or appeal against, the refusal and the authority to which such request for review or appeal shall lie, with the indication, where applicable, that the request for review or the appeal has to be filed through the intermediary of a representative whose address is within the territory of the Contracting Party whose Office has pronounced the refusal, and

(vii) the date on which the refusal was pronounced.

(3) [*Notification of Division of International Registration*] Where, following a notification of refusal in accordance with Article 13(2) of the 1999 Act, an international registration is divided before the Office of a designated Contracting Party in order to overcome a ground of refusal stated in that notification, that Office shall notify the International Bureau of such data concerning the division as shall be specified in the Administrative Instructions.

(4) [*Notification of Withdrawal of Refusal*]

(a) The notification of any withdrawal of refusal shall relate to one international registration, shall be dated and shall be signed by the Office making the notification.

(b) The notification shall contain or indicate

(i) the Office making the notification,

(ii) the number of the international registration,

(iii) where the withdrawal does not relate to all the industrial designs to which the refusal applied, those to which it relates or does not relate, and

(iv) the date on which the refusal was withdrawn.

(5) [*Recording*] The International Bureau shall record any notification received under paragraph (1)(c)(ii), (2) or (4) in the International Register together with, in the case of a notification of refusal, an indication of the date on which the notification of refusal was sent to the International Bureau.

(6) [*Transmittal of Copies of Notifications*] The International Bureau shall transmit copies of notifications received under paragraph (1)(c)(ii), (2) or (4) to the holder.

Rule 19
Irregular Refusals

(1) [*Notification Not Regarded as Such*]

(a) A notification of refusal shall not be regarded as such by the International Bureau and shall not be recorded in the International Register

(i) if it does not indicate the number of the international registration concerned, unless other indications contained in the notification permit the said registration to be identified,

(ii) if it does not indicate any grounds for refusal, or

(iii) if it is sent to the International Bureau after the expiry of the period applicable under Rule 18(1).

(b) Where subparagraph (a) applies, the International Bureau shall, unless it cannot identify the international registration concerned, transmit a copy of the notification to the holder, shall inform, at the same time, the holder and the Office that sent the notification that the notification of refusal is not regarded as such by the International Bureau and has not been recorded in the International Register, and shall indicate the reasons therefor.

(2) [*Irregular Notification*] If the notification of refusal

(i) is not signed on behalf of the Office which communicated the refusal, or does not comply with the requirements established under Rule 2,

(ii) does not comply, where applicable, with the requirements of Rule 18(2)(b)(iv),

(iii) does not indicate, where applicable, the authority to which a request for review or an appeal lies and the applicable time limit, reasonable under the circumstances, for lodging such a request or appeal (Rule 18(2)(b)(vi)),

(iv) does not indicate the date on which the refusal was pronounced (Rule 18(2)(b)(vii)),

the International Bureau shall nevertheless record the refusal in the International Register and transmit a copy of the notification to the holder. If so requested by the holder, the International Bureau shall invite the Office which communicated the refusal to rectify its notification without delay.

Rule 20
Invalidation in Designated Contracting Parties

(1) [*Contents of the Notification of Invalidation*] Where the effects of an international registration are invalidated in a designated Contracting Party and the invalidation is no longer subject to any review or appeal, the Office of the Contracting Party whose competent authority has pronounced the invalidation shall, where it is aware of the invalidation, notify the International Bureau accordingly. The notification shall indicate

(i) the authority which pronounced the invalidation,

(ii) the fact that the invalidation is no longer subject to appeal,

(iii) the number of the international registration,

(iv) where the invalidation does not relate to all the industrial designs that are the subject of the international registration, those to which it relates or does not relate,

(v) the date on which the invalidation was pronounced and its effective date.

(2) [*Recording of the Invalidation*] The International Bureau shall record the invalidation in the International Register, together with the data contained in the notification of invalidation.

Chapter 4
Changes and Corrections

Rule 21
Recording of a Change

(1) [*Presentation of the Request*]

(a) A request for the recording shall be presented to the International Bureau on the relevant official form where the request relates to any of the following:

(i) a change in the ownership of the international registration in respect of all or some of the industrial designs that are the subject of the international registration;

(ii) a change in the name or address of the holder;

(iii) a renunciation of the international registration in respect of any or all of the designated Contracting Parties;

(iv) a limitation, in respect of any or all of the designated Contracting Parties, to one or some of the industrial designs that are the subject of the international registration.

(b) The request shall be presented by the holder and signed by the holder; however, a request for the recording of a change in ownership may be presented by the new owner, provided that it is

(i) signed by the holder, or

(ii) signed by the new owner and accompanied by an attestation from the competent authority of the holder's Contracting Party that the new owner appears to be the successor in title of the holder.

(2) [*Contents of the Request*] The request for the recording of a change shall, in addition to the requested change, contain or indicate

(i) the number of the international registration concerned,

(ii) the name of the holder, unless the change relates to the name or address of the representative,

(iii) in case of a change in the ownership of the international registration, the name and address, given in accordance with the Administrative Instructions, of the new owner of the international registration,

(iv) in case of a change in the ownership of the international registration, the Contracting Party or Parties in respect of which the new owner fulfills the conditions to be the holder of an international registration,

(v) in case of a change in the ownership of the international registration that does not relate to all the industrial designs and to all the Contracting Parties, the numbers of the industrial designs and the designated Contracting Parties to which the change in ownership relates, and

(vi) the amount of the fees being paid and the method of payment, or instruction to debit the required amount of fees to an account opened with the International Bureau, and the identification of the party effecting the payment or giving the instructions.

(3) [*Request Not Admissible*] A change in the ownership of an international registration may not be recorded in respect of a designated Contracting Party if that Contracting Party is not bound by an Act to which the Contracting Party, or one of the Contracting Parties, indicated under paragraph (2)(iv) is bound.

(4) [*Irregular Request*] If the request does not comply with the applicable requirements, the International Bureau shall notify that fact to the holder and, if the request was made by a person claiming to be the new owner, to that person.

(5) [*Time Allowed to Remedy Irregularity*] The irregularity may be remedied within three months from the date of the notification of the irregularity by the International Bureau. If the irregularity is not remedied within the said three months, the request shall be considered abandoned and the International Bureau shall notify accordingly and at the same time the holder and, if the request was presented by a person claiming to be the new owner, that person, and shall refund any fees paid, after deduction of an amount corresponding to one-half of the relevant fees.

(6) [*Recording and Notification of a Change*]

(a) The International Bureau shall, provided that the request is in order, promptly record the change in the International Register and shall inform the holder. In the case of a recording of a change in ownership, the International Bureau will inform both the new holder and the previous holder.

(b) The change shall be recorded as of the date of receipt by the International Bureau of the request complying with the applicable requirements. Where however the request indicates that the change should be recorded after another change, or after renewal of the international registration, the International Bureau shall proceed accordingly.

(7) [*Recording of Partial Change in Ownership*] Assignment or other transfer of the international registration in respect of some only of the industrial designs, or some only of the designated Contracting Parties shall be recorded in the International Register under the number of the international registration of which a part has been assigned or otherwise transferred; any assigned or otherwise transferred part shall be canceled under the number of the said international registration and recorded as a separate international registration. The separate international registration shall bear the number of the international registration of which a part has been assigned or otherwise transferred, together with a capital letter.

(8) [*Recording of Merger of International Registrations*] Where the same person becomes the holder of two or more international registrations resulting from a partial change in ownership, the registrations shall be merged at the request of the said person and paragraphs (1) to (6) shall apply *mutatis mutandis*. The international registration resulting from the merger shall bear the number of the international registration of which a part had been assigned or otherwise transferred, together, where applicable, with a capital letter.

Rule 22
Corrections in the International Register

(1) [*Correction*] Where the International Bureau, acting *ex officio* or at the request of the holder, considers that there is an error concerning an international registration

in the International Register, it shall modify the Register and inform the holder accordingly.

(2) [*Refusal of Effects of Correction*] The Office of any designated Contracting Party shall have the right to declare in a notification to the International Bureau that it refuses to recognize the effects of the correction. Rules 18 and 19 shall apply *mutatis mutandis*.

Chapter 5
Renewals

Rule 23
Unofficial Notice of Expiry

Six months before the expiry of a five-year term, the International Bureau shall send to the holder and the representative, if any, a notice indicating the date of expiry of the international registration. The fact that the said notice is not received shall not constitute an excuse for failure to comply with any time limit under Rule 24.

Rule 24
Details Concerning Renewal

(1) [*Fees*]

(a) The international registration shall be renewed upon payment of the following fees:

(i) a basic fee;

(ii) a standard designation fee in respect of each Contracting Party designated under the 1999 Act that has not made a declaration under Article 7(2) of the 1999 Act, and each Contracting Party designated under the 1960 Act, for which the international registration is to be renewed;

(iii) an individual designation fee for each Contracting Party designated under the 1999 Act that has made a declaration under Article 7(2) of the 1999 Act and for which the international registration is to be renewed.

(b) The amounts of the fees referred to in items (i) and (ii) of subparagraph (a) are set out in the Schedule of Fees.

(c) The payment of the fees referred to in subparagraph (a) shall be made at the latest on the date on which the renewal of the international registration is due. However, it may still be made within six months from the date on which the renewal of the international registration is due, provided that the surcharge specified in the Schedule of Fees is paid at the same time.

(d) If any payment made for the purposes of renewal is received by the International Bureau earlier than three months before the date on which the renewal of the international registration is due, it shall be considered as having been received three months before that date.

(2) [*Further Details*]

(a) Where the holder does not wish to renew the international registration

(i) in respect of a designated Contracting Party, or

(ii) in respect of any of the industrial designs that are the subject of the international registration,

payment of the required fees shall be accompanied by a statement indicating the Contracting Party or the numbers of the industrial designs for which the international registration is not to be renewed.

(b) Where the holder wishes to renew the international registration in respect of a designated Contracting Party notwithstanding the fact that the maximum period of protection for industrial designs in that Contracting Party has expired, payment of the required fees, including the standard designation fee or the individual designation fee, as the case may be, for that Contracting Party, shall be accompanied by a statement that the renewal of the international registration is to be recorded in the International Register in respect of that Contracting Party.

(c) Where the holder wishes to renew the international registration in respect of a designated Contracting Party notwithstanding the fact that a refusal is recorded in the International Register for that Contracting Party in respect of all the industrial designs concerned, payment of the required fees, including the standard designation fee or the individual designation fee, as the case may be, for that Contracting Party, shall be accompanied by a statement specifying that the renewal of the international registration is to be recorded in the International Register in respect of that Contracting Party.

(d) The international registration may not be renewed in respect of any designated Contracting Party in respect of which an invalidation has been recorded for all the industrial designs under Rule 20 or in respect of which a renunciation has been recorded under Rule 21. The international registration may not be renewed in respect of any designated Contracting Party for those industrial designs in respect of which an invalidation in that Contracting Party has been recorded under Rule 20 or in respect of which a limitation has been recorded under Rule 21.

(3) [*Insufficient Fees*]

(a) If the amount of the fees received is less than the amount required for renewal, the International Bureau shall promptly notify at the same time both the holder and the representative, if any, accordingly. The notification shall specify the missing amount.

(b) If the amount of the fees received is, on the expiry of the period of six months referred to in paragraph (1)(c), less than the amount required for renewal, the International Bureau shall not record the renewal, shall refund the amount received and shall notify accordingly the holder and the representative, if any.

Rule 25
Recording of the Renewal; Certificate

(1) [*Recording and Effective Date of the Renewal*] Renewal shall be recorded in the International Register with the date on which renewal was due, even if the fees required for renewal are paid within the period of grace referred to in Rule 24(1)(c).

(2) [*Certificate*] The International Bureau shall send a certificate of renewal to the holder.

Chapter 6
Bulletin

Rule 26
Bulletin

(1) [*Information Concerning International Registrations*] The International Bureau shall publish in the Bulletin relevant data concerning

(i) international registrations, in accordance with Rule 17;

(ii) refusals recorded under Rule 18(5), with an indication as to whether there is a possibility of review or appeal, but without the grounds for refusal;

(iii) invalidations recorded under Rule 20(2);

(iv) changes in ownership, changes of name or address of the holder, renunciations and limitations recorded under Rule 21;

(v) corrections effected under Rule 22;

(vi) renewals recorded under Rule 25(1);

(vii) international registrations which have not been renewed.

(2) [*Information Concerning Declarations; Other Information*] The International Bureau shall publish in the Bulletin any declaration made by a Contracting Party under the 1999 Act, the 1960 Act or these Regulations, as well as a list of the days on which the International Bureau is not scheduled to open to the public during the current and the following calendar year.

(3) [*Mode of Publishing the Bulletin*] The Bulletin referred to in Rule 1(1)(x) shall be published on the website of the Organization. The date on which each issue of the Bulletin is published on that website shall be electronically communicated by the International Bureau to the Office of each Contracting Party. Such communication shall be deemed to replace the sending of the Bulletin referred to in Article 10(3)(b) of the 1999 Act and Article 6(3)(b) of the 1960 Act, and, for the purposes of Article 8(2) of the 1960 Act, the Bulletin shall be deemed to have been received by each Office concerned also on the date of the said communication.

Chapter 7
Fees

Rule 27
Amounts and Payment of Fees

(1) [*Amounts of Fees*] The amounts of fees due under the 1999 Act, the 1960 Act, the 1934 Act and these Regulations, other than individual designation fees referred to in Rule 12(1)(a)(iii), shall be specified in the Schedule of Fees which is annexed to these Regulations and forms an integral part thereof.

(2) [*Payment*]

(a) Subject to subparagraph (b) and Rule 12(3)(c), the fees shall be paid directly to the International Bureau.

(b) Where the international application is filed through the Office of the applicant's Contracting Party, the fees payable in connection with that application may be paid through that Office if it accepts to collect and forward such fees and the applicant or the holder so wishes. Any Office which accepts to collect and forward such fees shall notify that fact to the Director General.

(3) [*Modes of Payment*] Fees shall be paid to the International Bureau in accordance with the Administrative Instructions.

(4) [*Indications Accompanying the Payment*] At the time of the payment of any fee to the International Bureau, an indication must be given,

(i) before international registration, of the name of the applicant, the industrial design concerned and the purpose of the payment;

(ii) after international registration, of the name of the holder, the number of the international registration concerned and the purpose of the payment.

(5) [*Date of Payment*]

(a) Subject to Rule 24(1)(d) and subparagraph (b), any fee shall be considered to have been paid to the International Bureau on the day on which the International Bureau receives the required amount.

(b) Where the required amount is available in an account opened with the International Bureau and that Bureau has received instructions from the holder of the account to debit it, the fee shall be considered to have been paid to the International Bureau on the day on which the International Bureau receives an international application, a request for the recording of a change, or an instruction to renew an international registration.

(6) [*Change in the Amount of the Fees*]

(a) Where an international application is filed through the Office of the applicant's Contracting Party and the amount of the fees payable in respect of the filing of the international application is changed between, on the one hand, the date on which the international application was received by that Office and, on the other hand, the date of the receipt of the international application by the International Bureau, the fee that was valid on the first date shall be applicable.

(b) Where the amount of the fees payable in respect of the renewal of an international registration is changed between the date of payment and the due date of the renewal, the fee that was valid on the date of payment, or on the date considered to be the date of payment under Rule 24(1)(d), shall be applicable. Where the payment is made after the due date, the fee that was valid on the due date shall be applicable.

(c) Where the amount of any fee other than the fees referred to in subparagraphs (a) and (b) is changed, the amount valid on the date on which the fee was received by the International Bureau shall be applicable.

Rule 28
Currency of Payments

(1) [*Obligation to Use Swiss Currency*] All payments made under these Regulations to the International Bureau shall be in Swiss currency irrespective of the fact that, where the fees are paid through an Office, such Office may have collected those fees in another currency.

(2) [*Establishment of the Amount of Individual Designation Fees in Swiss Currency*]

(a) Where a Contracting Party makes a declaration under Article 7(2) of the 1999 Act or under Rule 36(1) that it wants to receive an individual designation fee, the amount of the fee indicated to the International Bureau shall be expressed in the currency used by its Office.

(b) Where the fee is indicated in the declaration referred to in subparagraph (a) in a currency other than Swiss currency, the Director General shall, after consultation with the Office of the Contracting Party concerned, establish the amount of the fee in Swiss currency on the basis of the official exchange rate of the United Nations.

(c) Where, for more than three consecutive months, the official exchange rate of the United Nations between the Swiss currency and the currency in which the amount of an individual designation fee has been indicated by a Contracting Party is higher or lower by at least 5% than the last exchange rate applied to establish the amount of the fee in Swiss currency, the Office of that Contracting Party may ask the Director General to establish a new amount of the fee in Swiss currency according to the official exchange rate of the United Nations prevailing on the day preceding the day on which the request is made. The Director General shall proceed accordingly. The new amount shall be applicable as from a date which shall be fixed by the Director General, provided that such date is between one and two months after the date of the publication of the said amount in the Bulletin.

(d) Where, for more than three consecutive months, the official exchange rate of the United Nations between the Swiss currency and the currency in which the amount of an individual designation fee has been indicated by a Contracting Party is lower by at least 10% than the last exchange rate applied to establish the amount of the fee in Swiss currency, the Director General shall establish a new amount of the fee in Swiss currency according to the current official exchange rate of the United Nations. The new amount shall be applicable as from a date which shall be fixed by the Director General, provided that such date is between one and two months after the date of the publication of the said amount in the Bulletin.

Rule 29
Crediting of Fees to the Accounts of the Contracting Parties Concerned

Any standard designation fee or individual designation fee paid to the International Bureau in respect of a Contracting Party shall be credited to the account of that Contracting Party with the International Bureau within the month following the month in the course of which the recording of the international registration or renewal for which that fee has been paid was effected or, as regards the second part of the individual designation fee, immediately upon its receipt by the International Bureau.

Chapter 8
International Applications Governed Exclusively or Partly by the 1934 Act and International Registrations Resulting Therefrom

Rule 30
Applicability of These Regulations to International Applications Governed Exclusively by the 1934 Act and International Registrations Resulting Therefrom

(1) [*General Principle*] Unless otherwise specified and subject to paragraph (2), these Regulations shall be applicable to international applications governed exclusively by the 1934 Act and to international registrations resulting therefrom.

(2) [*Exceptions*]

(a) Notwithstanding Rule 6, any international application governed exclusively by the 1934 Act shall be in French. Any communication concerning that international application or the international registration resulting therefrom shall be in French.

(b) Notwithstanding Rule 7(5)(a), a description of the reproduction or of the characteristic features of the industrial designs and the identity of the creator of the industrial designs may not be included in an application governed exclusively by the 1934 Act.

(c) Notwithstanding Rule 7(5)(e), deferment of publication may not be requested with respect to an international application governed exclusively by the 1934 Act.

(d) Notwithstanding Rule 7(7), the industrial designs contained in an international application governed exclusively by the 1934 Act may belong to different classes of the International Classification.

(e) Notwithstanding Rule 9(1), an international application governed exclusively by the 1934 Act may, instead of containing reproductions, be accompanied by specimens.

(f) Notwithstanding Rule 12(1)(a), an international application governed exclusively by the 1934 Act shall be subject only to the payment of the basic fee referred to in Rule 12(1)(a)(i).

(g) Notwithstanding Rule 15(1), the reproductions of the industrial designs contained in an international application governed exclusively by the 1934 Act shall not be registered in the International Register.

(h) Notwithstanding Rule 17(1), an international registration resulting from an international application governed exclusively by the 1934 Act shall be published immediately after registration.

(i) Notwithstanding Rule 17(2)(ii), the reproductions of industrial designs contained in an international registration resulting from an international application governed exclusively by the 1934 Act shall not be published in the Bulletin.

(j) Notwithstanding Rule 18, the effects of an international registration resulting from an international application governed exclusively by the 1934 Act may not be the subject of a notification of refusal of protection.

(k) Notwithstanding Rule 21(3), a change in ownership may not be recorded in respect of a Contracting Party designated under the 1934 Act if, on the basis of the indications referred to in Rule 21(2)(iv), the 1934 Act would cease to be applicable in respect of that Contracting Party.

(l) Notwithstanding Rule 24(1)(a), an international registration resulting from an international application governed exclusively by the 1934 Act shall be renewed upon payment only of the basic fee referred to in Rule 24(1)(a)(i).

(m) Notwithstanding Rule 24(2)(b), an international registration resulting from an international application governed exclusively by the 1934 Act shall not be renewed where the duration of international protection of fifteen years referred to in Article 7 of the 1934 Act has expired.

(n) The renewal of an international registration resulting from an international application governed exclusively by the 1934 Act, for the second period of protection of ten years referred to in Article 7 of the 1934 Act, may be requested at the time of filing the international application concerned. In such case, the fee referred to in Rule 24(1)(a)(i) shall be payable at the time of filing of the said international application, failing which the request for renewal shall be disregarded by the International Bureau.

(3) [*International Applications Under Sealed Cover*]

(a) An international application governed exclusively by the 1934 Act shall contain, in addition to the indications referred to in Rule 7(3), an indication as to whether such application is made under open or sealed cover.

(b) Upon expiry of the first period of five years of international protection referred to in Article 7 of the 1934 Act, any international registration which is under sealed cover shall be opened by the International Bureau upon renewal of the said registration.

Rule 31
Applicability of These Regulations to International Applications Governed Partly by the 1934 Act and International Registrations Resulting Therefrom

(1) [*General Principle*] These Regulations shall be applicable to international applications referred to in Rule 1(1)(xviii) to (xx), and to international registrations resulting therefrom, subject to paragraph (2).

(2) [*Exceptions*]

(a) Notwithstanding Rule 7(5)(e), deferment of publication may not be requested with respect to an international application referred to in paragraph (1). Where deferment of publication has been requested and one of the designated Contracting Parties in the international application has been designated under the 1934 Act, the International Bureau shall notify the applicant accordingly; if, within the period of one month from the date of the notification sent by the International Bureau, the applicant does not, by notice in writing to the International Bureau, withdraw the designation of the said Contracting Party, the International Bureau shall disregard the request for deferment of publication.

(b) Notwithstanding Rule 21(3), a change in ownership may not be recorded in respect of a designated Contracting Party if, on the basis of the indications referred to in Rule 21(2)(iv), the 1934 Act would cease to be applicable, or would become applicable, in respect of that Contracting Party.

(c) With respect to Contracting Parties designated under the 1934 Act in an international application referred to in paragraph (1), or in an international registration resulting therefrom,

(i) the fees referred to in Rule 12(1)(a)(ii) to (iv) shall not be payable;
(ii) the effects of the international registration concerned may not be the subject of a notification of refusal of protection referred to in Rule 18;
(iii) the renewal fees referred to in Rule 24(1)(a)(ii) and (iii) shall not be payable;

(iv) the international registration concerned shall not be renewed where the duration of international protection of fifteen years referred to in Article 7 of the 1934 Act has expired, notwithstanding Rule 24(2)(b).

Chapter 9
Miscellaneous

Rule 32
Extracts, Copies and Information Concerning Published International Registrations

(1) [*Modalities*] Against payment of a fee whose amount shall be fixed in the Schedule of Fees, any person may obtain from the International Bureau, in respect of any published international registration:

(i) extracts from the International Register;

(ii) certified copies of recordings made in the International Register or of items in the file of the international registration;

(iii) uncertified copies of recordings made in the International Register or of items in the file of the international registration;

(iv) written information on the contents of the International Register or of the file of the international registration;

(v) a photograph of a specimen.

(2) [*Exemption from Authentication, Legalization or any Other Certification*] In respect of a document referred to in paragraph (1)(i) and (ii), bearing the seal of the International Bureau and the signature of the Director General or a person acting on his behalf, no authority of any Contracting Party shall require authentication, legalization or any other certification of such document, seal or signature, by any other person or authority. The present paragraph applies *mutatis mutandis* to the international registration certificate referred to in Rule 15(1).

Rule 33
Amendment of Certain Rules

(1) [*Requirement of Unanimity*] Amendment of the following provisions of these Regulations shall require unanimity of the Contracting Parties bound by the 1999 Act:

(i) Rule 13(4);

(ii) Rule 18(1).

(2) [*Requirement of Four-Fifths Majority*] Amendment of the following provisions of the Regulations and of paragraph (3) of the present Rule shall require a four-fifths majority of the Contracting Parties bound by the 1999 Act:

(i) Rule 7(7);

(ii) Rule 9(3)(b);

(iii) Rule 16(1)(a);

(iv) Rule 17(1)(iii).

(3) [*Procedure*] Any proposal for amending a provision referred to in paragraph (1) or (2) shall be sent to all Contracting Parties at least two months prior to the opening of the session of the Assembly which is called upon to make a decision on the proposal.

Rule 34
Administrative Instructions

(1) [*Establishment of Administrative Instructions; Matters Governed by Them*]

(a) The Director General shall establish Administrative Instructions. The Director General may modify them. The Director General shall consult the Offices of the Contracting Parties with respect to the proposed Administrative Instructions or their proposed modification.

(b) The Administrative Instructions shall deal with matters in respect of which these Regulations expressly refer to such Instructions and with details in respect of the application of these Regulations.

(2) [*Control by the Assembly*] The Assembly may invite the Director General to modify any provision of the Administrative Instructions, and the Director General shall proceed accordingly.

(3) [*Publication and Effective Date*]

(a) The Administrative Instructions and any modification thereof shall be published in the Bulletin.

(b) Each publication shall specify the date on which the published provisions become effective. The dates may be different for different provisions, provided that no provision may be declared effective prior to its publication in the Bulletin.

(4) [*Conflict with the 1999 Act, the 1960 Act, the 1934 Act or These Regulations*] In the case of conflict between, on the one hand, any provision of the Administrative

Instructions and, on the other hand, any provision of the 1999 Act, the 1960 Act, the 1934 Act or of these Regulations, the latter shall prevail.

Rule 35
Declarations Made by Contracting Parties to the 1999 Act

(1) [*Making and Coming into Effect of Declarations*] Article 30(1) and (2) of the 1999 Act shall apply *mutatis mutandis* to the making of any declaration under Rules 8(1), 9(3)(a), 13(4) or 18(1)(b) and to its coming into effect.

(2) [*Withdrawal of Declarations*] Any declaration referred to in paragraph (1) may be withdrawn at any time by notification addressed to the Director General. Such withdrawal shall take effect upon receipt by the Director General of the notification of withdrawal or at any later date indicated in the notification. In the case of a declaration made under Rule 18(1)(b), the withdrawal shall not affect an international registration whose date is earlier than the coming into effect of the said withdrawal.

Rule 36
Declarations Made by Contracting Parties to the 1960 Act

(1) [*Individual Designation Fee*] For the purpose of Article 15(1)2(b) of the 1960 Act, any Contracting Party to the 1960 Act whose Office is an Examining Office may, in a declaration, notify the Director General that, in connection with any international application in which it is designated under the 1960 Act, the standard designation fee referred to in Rule 12(1)(a)(ii) shall be replaced by an individual designation fee, whose amount shall be indicated in the declaration and can be changed in further declarations. The said amount may not be higher than the equivalent of the amount which the Office of that Contracting Party would be entitled to receive from an applicant for a grant of protection for an equivalent period to the same number of industrial designs, that amount being diminished by the savings resulting from the international procedure.

(2) [*Maximum Duration of Protection*] Each Contracting Party to the 1960 Act shall, in a declaration, notify the Director General of the maximum duration of protection provided for by its law.

(3) [*Time at Which Declarations May Be Made*] Any declaration under paragraphs (1) and (2) may be made

(i) at the time of the deposit of an instrument referred to in Article 26(2) of the 1960 Act, in which case it shall become effective on the date on which the State having made the declaration becomes bound by this Act, or

(ii) after the deposit of an instrument referred to in Article 26(2) of the 1960 Act, in which case it shall become effective one month after the date of its receipt by the Director General or at any later date indicated in the declaration but shall apply only in respect of any international registration whose date of international registration is the same as, or is later than, the effective date of the declaration.

Schedule of Fees (as in force on April 1, 2004)

I. International Applications Governed Exclusively or Partly by the 1960 Act or by the 1999 Act

			Swiss francs
1.	Basic fee		
	1.1	For one design	397
	1.2	For each additional design included in the same international application	19
2.	Publication fee		
	2.1	For each reproduction to be published in black and white	12
	2.2	For each reproduction to be published in color	75
	2.3	For each page, in addition to the first, on which one or more reproductions are shown (where the reproductions are submitted on paper)	150
3.	Additional fee where the description exceeds 100 words per word exceeding 100 words		2
4.	Standard designation fee		
	4.1	For one design	42
	4.2	For each additional design included in the same international application	2
5.	Individual designation fee (the amount of the individual designation fee is fixed by each Contracting Party concerned)		

II. International Applications Governed Exclusively by the 1934 Act

6.	Basic fee		
	6.1	For one design	216
	6.2	For two to 50 designs included in the same international application	432

6.3	For 51 to 100 designs included in the same international application	638

III. Renewal of an International Registration Resulting From an International Application Governed Exclusively or Partly by the 1960 Act or by the 1999 Act

7.	Basic fee	
7.1	For one design	200
7.2	For each additional design included in the same international registration	17
8.	Standard designation fee	
8.1	For one design	21
8.2	For each additional design included in the same international registration	1
9.	Individual designation fee (the amount of the individual designation fee is fixed by each Contracting Party concerned)	
10.	Surcharge (period of grace)	**

IV. Renewal of an International Registration Resulting From an International Application Governed Exclusively by the 1934 Act

11.	Basic fee	
11.1	For one design	422
11.2	For two to 50 designs included in the same international registration	844
11.3	For 51 to 100 designs included in the same international registration	1236
12.	Surcharge (period of grace)	**

V. Miscellaneous Recordings

13.	Change in ownership	144
14.	Change of name and/or address of the holder	
14.1	For one international registration	144
14.2	For each additional international registration of the same holder included in the same request	72

** 50% of the renewal basic fee.

15.	Renunciation	144
16.	Limitation	144

VI. Information Concerning Published International Registrations

17.	Supply of an extract from the International Register relating to a published international registration	144
18.	Supply of non-certified copies of the International Register or of items in the file of a published international registration	
	18.1 For the first five pages	26
	18.2 For each additional page after the fifth if the copies are requested at the same time and relate to the same international registration	2
19.	Supply of certified copies from the International Register or of items in the file of a published international registration	
	19.1 For the first five pages	46
	19.2 For each additional page after the fifth if the copies are requested at the same time and relate to the same international registration	2
20.	Supply of a photograph of a specimen	57
21.	Supply of written information on the contents of the International Register or of the file of a published international registration	
	21.1 Concerning one international registration	82
	21.2 Concerning any additional international registration of the same holder if the same information is requested at the same time	10
22.	Search in the list of owners of international registrations	
	22.1 Per search by the name of a given person or entity	82
	22.2 For each international registration found beyond the first one	10
23.	Surcharge for the communication of extracts, copies, information or search reports by telefacsimile (per page)	4

APPENDIX III

Administrative Instructions for the Application of the Hague Agreement (as in force on April 1, 2004)

Table of Contents

Part One
Definitions

Section 101
Abbreviated Expressions

(a) For the purposes of these Administrative Instructions:

(i) "Regulations" means the Common Regulations under the Hague Agreement Concerning the International Registration of Industrial Designs;

(ii) "Rule" means a Rule of the Regulations;

(b) An expression which is used in these Administrative Instructions and is referred to in Rule 1 has the same meaning as in the Regulations.

Part Two
Communications with the International Bureau

Section 201
Communication in Writing; Several Documents in One Envelope

(a) Communications addressed to the International Bureau shall be effected in writing by typewriter or other machine and shall be signed.

(b) If several documents are mailed in one envelope, they should be accompanied by a list identifying each of them.

Section 202
Signature

A signature shall be hand written, printed or stamped; it may be replaced by the affixing of a seal or, as regards the electronic communication referred to in Section 204, by a mode of identification to be determined by the International Bureau.

Section 203
Communication by Telefacsimile

(a) Any communication, other than an international application containing a reproduction to be published in color, may be addressed to the International Bureau by telefacsimile, provided that where the communication must be presented on an official form, the official form is used for the purposes of the telefacsimile communication.

(b) An international application addressed to the International Bureau by telefacsimile shall not have effect unless, prior to the expiration of 20 days counted from the date of receipt of such communication, the original of the international application bearing the prescribed signature along with the reproductions and/or specimens concerned, have been received by the International Bureau. When so confirmed, the said international application shall have effect as of the date on which it was received by telefacsimile by the International Bureau.

(c) Where an international application has been transmitted to the International Bureau by telefacsimile, the Bureau shall promptly and by telefacsimile inform the sender of the receipt of that international application, and where such telefacsimile communication received by the International Bureau is incomplete or illegible, of that fact also, provided that the sender can be identified and can be reached by telefacsimile.

Section 204
Electronic Communications

(a) Communications with the International Bureau, including the presentation of the international application, may take place by electronic means at a time and in a manner and format to be ascertained by the International Bureau, the particulars of which shall be published in the Bulletin.

(b) The International Bureau shall promptly and by electronic transmission inform the originator of an electronic transmission of the receipt of that transmission and, where the electronic transmission received is incomplete or otherwise unusable, also of that fact, provided that the originator can be identified and can be reached.

(c) For the purpose of the communication by the International Bureau to Offices of Contracting Parties of the date on which each issue of the Bulletin is published, as provided for in Rule 26(3), each such Office shall indicate to the International Bureau the email address to which the said communication shall be sent.

Part Three
Requirements Concerning Names and Addresses

Section 301
Names and Addresses

(a) In the case of a natural person, the name to be indicated is the family or principal name and the given or secondary name(s) of the natural person.

(b) In the case of a legal entity, the name to be indicated is the full official designation of the legal entity.

(c) In the case of a name in characters other than Latin characters, the indication of that name shall consist of a transliteration into Latin characters which shall follow the phonetics of the language of the international application. In the case of a legal entity whose name is in characters other than Latin characters, the said transliteration may be replaced by a translation into the language of the international application.

(d) An address shall be given in such a way as to satisfy the customary requirements for prompt postal delivery and shall consist, at least, of all the relevant administrative units up to, and including, the house number, if any. In addition, telephone and telefacsimile numbers, an e-mail address as well as a different address for correspondence, may be indicated.

Section 302
Address for Correspondence

Where there are two or more applicants or new owners with different addresses and no representative is appointed, one address for correspondence shall be indicated. Where no such address has been indicated, the address of the person named first shall be treated as the address for correspondence.

Part Four
Reproduction of the Industrial Design; Disclaimer; Numbering; Description of Characteristic Features

Section 401
Presentation of Reproductions

(a) One and the same international application may comprise both photographs and other graphic representations, in black and white or in color.

(b) Each reproduction accompanying an international application shall be submitted in a single copy.

(c) The photographs or other graphic representations accompanying the international application shall be either pasted or printed directly onto a separate sheet of A4 paper which is white and opaque. The separate sheet of paper shall be used upright and shall not contain more than 25 reproductions.

(d) The reproductions must be arranged on the separate sheet referred to in sub-paragraph (c) in the orientation in which the applicant wishes them to be published. A margin of at least 5 millimeters should be left around the representation of each industrial design.

(e) Each reproduction must fall within a right-angled quadrilateral containing no other reproduction or part of another reproduction and no numbering. The photographs or other graphic representations shall not be folded, stapled or marked in any way.

Section 402
Representation of the Industrial Design

(a) The photographs and other graphic representations shall represent the industrial design alone, or the product in relation to which the industrial design is to be used, to the exclusion of any other object, accessory, person or animal.

(b) The dimensions of the representation of each industrial design appearing in a photograph or other graphic representation may not exceed 16 x 16 centimeters, and one of those dimensions must be at least 3 centimeters.

(c) The following shall not be accepted:

(i) technical drawings, particularly with axes and dimensions;

(ii) explanatory text or legends.

Section 403
Disclaimer

Matter which is shown in a reproduction but for which protection is not sought may be indicated

(i) in the description referred to in Rule 7(5)(a) and/or

(ii) by means of dotted or broken lines.

Section 404
Requirements for Photographs and Other Graphic Representations

(a) The photographs supplied must be of professional standard and have all the edges cut at right angles. The industrial design must be shown against a neutral plain background. Photographs retouched with ink or correcting fluid shall not be allowed.

(b) Graphic representations must be of professional standard produced with drawing instruments or by electronic means on good quality white, opaque paper, all of whose edges are cut at right angles. The industrial design represented may comprise shading and hatching to provide relief. Graphic representations executed by electronic means may be shown against a background, provided that it is neutral and plain and has only edges cut at right angles.

Section 405
Numbering of Reproductions

(a) The numbering stipulated for multiple international applications shall appear in the margin of each photograph or other graphic representation. When the same industrial design is represented from different angles, the numbering shall consist of two separate figures separated by a dot (e.g., 1.1, 1.2, 1.3, etc. for the first design, 2.1, 2.2, 2.3, etc. for the second design, and so on).

(b) The reproductions shall be pasted or printed in ascending numerical order on the separate sheet referred to in Section 401(c).

Section 406
Requirements for Specimens

(a) A specimen in two dimensions which accompanies an international application shall not exceed 26.2 centimeters x 17 centimeters in size (unfolded), 50 grams in weight or 3 millimeters in thickness. Such specimens shall be pasted on sheets of A4 paper and numbered in accordance with Section 405(b). The same number shall be assigned to each reproduction corresponding to these specimens when submitted to the International Bureau.

(b) None of the dimensions of a package containing specimens shall exceed 30 centimeters and the weight of such package and its packing shall not exceed 4 kilograms.

(c) Perishable products or products which may be dangerous to store shall not be accepted.

Part Five
Refusals

Section 501
Date of Sending of Notification of Refusal

In the case of a notification of refusal sent through a postal service, the date of dispatch shall be determined by the postmark. If the postmark is illegible or missing, the International Bureau shall treat such notification as if it had been sent 20 days before the date of its receipt by the International Bureau. However, if the date of dispatch thus determined is earlier than any date of refusal or date of sending mentioned in the notification, the International Bureau shall treat such notification as if it had been sent on the latter date. In the case of a notification of refusal sent through a delivery service, the date of dispatch shall be determined by the indication given by such delivery service on the basis of the details of the mailing as recorded by it.

Section 502
Notification of Division of an International Registration

Where an international registration has been divided before the Office of a designated Contracting Party following a notification of refusal as provided for in Rule 18(3), that Office shall notify the International Bureau of that fact, together with the following additional particulars:

(i) the Office making the notification;

(ii) the number of the international registration concerned;

(iii) the numbers of the industrial designs which have been the subject of the division with the Office concerned, and

(iv) the resulting national or regional application numbers or registration numbers.

Part Six
Request for the Recording of a Limitation or Renunciation when Publication Deferred

Section 601
Latest Time to Request the Recording of a Limitation or Renunciation

Where the publication of an international registration is deferred, a request for the recording of a limitation or renunciation concerning that registration, complying with the applicable requirements, must be received by the International Bureau not later than three months prior to the expiry of the period of deferment. In default of this, the international registration is published at the expiration of the period of deferment without account being taken of the request for the recording of a limitation or renunciation. Provided that the request for limitation or renunciation complies with the applicable requirements, the limitation or renunciation is however recorded in the International Register.

Part Seven
Renewal

Section 701
Unofficial Notice of Expiry

When, pursuant to Rule 23, the International Bureau sends to the holder and the representative, if any, a notice indicating the date of expiration of an international registration, such notice shall contain also an indication of the Contracting Parties for which, at the date of the notice, and in accordance with the maximum duration of protection notified by each Contracting Party pursuant to Article 17(3)(c) of the 1999 Act and Rule 36(2), renewal of the international registration is possible.

Part Eight
Fees

Section 801
Mode of Payment

Fees may be paid to the International Bureau

(i) by debit to a current account with the International Bureau,

(ii) by payment into the Swiss postal check account or to any of the specified bank accounts of the International Bureau,

(iii) by a banker's check,

(iv) by payment in cash at the International Bureau.

APPENDIX IV

London Act of June 2, 1934

Table of Contents

Article 1

Nationals of any of the contracting countries, as well as persons who, upon the territory of the restricted Union, have satisfied the conditions of Article 3 of the General Convention, may, in all the other contracting countries, secure protection for their industrial designs by means of an international deposit made at the International Bureau of Industrial Property at Berne.

Article 2

(1) The international deposit shall include designs, either in the form of the industrial product for which they are intended, or in the form of a drawing, a photograph, or any other adequate graphic representation of the said design.

(2) The articles submitted shall be accompanied by an application for international deposit, in duplicate, containing, in French, the particulars specified in the Regulations.

Article 3

(1) As soon as the International Bureau receives an application for international deposit it shall record the application in a special Register and shall publish it, sending free of charge to each Office the desired number of copies of the periodical journal in which it publishes such records.

(2) Deposits shall be kept in the archives of the International Bureau.

Article 4

(1) Any person making an international deposit of an industrial design shall, in the absence of proof to the contrary, be deemed to be the owner of the work.

(2) International deposit is purely declaratory. The deposit, as such, shall have in the contracting countries the same effect as if the designs had been deposited there direct on the date of the international deposit, subject however to the special rules established by this Agreement.

(3) The publicity mentioned in the foregoing Article shall be deemed in all the contracting countries to be sufficient, and no other publicity may be required of the depositor, apart from any formalities to be complied with for the exercise of rights in accordance with the domestic law.

(4) The right of priority established by Article 4 of the General Convention shall be guaranteed to every design which has been the subject of an international deposit, without requiring compliance with any of the formalities prescribed in the said Article.

Article 5

The contracting countries agree not to require designs which have been the subject of an international deposit to bear any compulsory marking. They shall not cause the designs to lapse either by reason of non-exploitation or as a result of the introduction of articles similar to those protected.

Article 6

(1) The international deposit may consist of either a single design or several, the number thereof being stated in the application.

(2) Such deposit may be made under open cover or under sealed cover. In particular, there shall be accepted as a means of deposit under sealed cover double envelopes having a perforated control number (Soleau system) or any other system for ensuring identification.

(3) The maximum dimensions of covers or packets eligible for deposit shall be fixed by the Regulations.

Article 7

The duration of international protection is fixed at fifteen years from the date of deposit at the International Bureau at Berne; this term is divided into two periods, namely, one period of five years and one period of ten years.

Article 8

During the first period of protection, deposits shall be accepted either under open cover or under sealed cover; during the second period of protection, they shall be accepted only under open cover.

Article 9

During the first period, deposits under sealed cover may be opened at the request of the depositor or of a competent court; upon expiration of the first period, they shall, with a view to transition to the second period, be opened upon application for prolongation.

Article 10

In the course of the first six months of the fifth year of the first period, the International Bureau shall give unofficial notice of expiration to the depositor of the design.

Article 11

(1) When the depositor desires to secure extension of the protection by transition to the second period he shall, before the expiration of the first period, file with the International Bureau an application for prolongation.

(2) The International Bureau shall open the package, if sealed, shall publish in its journal notice of the prolongation granted, and shall notify all Offices thereof by sending to each the desired number of copies of the said journal.

Article 12

Designs forming the subject of deposits which have not been prolonged, as well as those in respect of which protection has expired, shall, upon the request of the proprietors and at their expense, be returned to them as they stand. If unclaimed, they shall be destroyed at the end of two years.

Article 13

(1) Depositors may, at any time, renounce their deposit, either wholly or in part, by means of a declaration addressed to the International Bureau; the Bureau shall give such declaration the publicity referred to in Article 3.

(2) Renunciation shall entail the return of the deposit to the depositor, at his expense.

Article 14

If a court or any other competent authority orders the communication to it of a secret design, the International Bureau, when duly required, shall open the deposited package, shall extract therefrom the requested design, and shall remit it to the authority so requiring. Similar communication shall take place on request in the case of an unsealed design. The article thus communicated shall be returned in the shortest possible time and reincorporated in the sealed package or in the envelope, as the case may be. Such transactions may be subject to a fee, which shall be fixed by the Regulations.

Article 15

The fees for an international deposit and for the prolongation thereof, which shall be paid before registration of the deposit or of the prolongation can be proceeded with, shall be as follows:

1. for a single design, and in respect of the first period of five years: 5 francs;
2. for a single design, upon expiration of the first period and in respect of the duration of the second period of ten years: 10 francs;
3. for a multiple deposit, and in respect of the first period of five years: 10 francs;

4. for a multiple deposit, upon expiration of the first period and in respect of the duration of the second period of ten years: 50 francs.

Article 16

The net annual proceeds from fees shall be divided, as provided in Article 8 of the Regulations, among the contracting countries by the International Bureau, after deduction of the common expenses necessitated by the implementation of this Agreement.

Article 17

(1) The International Bureau shall record in its Registers all changes affecting the proprietorship of designs which are notified to it by the parties concerned; it shall publish them in its journal and shall announce them to all Offices by sending to each the desired number of copies of the said journal.

(2) These transactions may be subject to a fee, which shall be fixed by the Regulations.

(3) The proprietor of an international deposit may assign the rights in respect of part only of the designs included in a multiple deposit or in respect of one or several of the contracting countries only; but, in such cases, if the deposit has been made under sealed cover, the International Bureau shall open the package before recording the transfer in its Registers.

Article 18

(1) The International Bureau shall deliver to any person, upon application, and on payment of a fee fixed by the Regulations, an abstract of the entries in the Register in connection with any given design.

(2) Such abstract may, if the design lends itself thereto, be accompanied by a copy or a reproduction of the design, which has been supplied to the International Bureau and which the latter shall certify as being in conformity with the article deposited under open cover. If the Bureau is not in possession of such copies or reproductions, it shall have them made, on the request of interested parties and at their expense.

Article 19

The archives of the International Bureau, in so far as they contain unsealed deposits, shall be accessible to the public. Any person may inspect them, in the presence of an official, or may obtain from the Bureau written information on the contents of the Register, subject to payment of fees to be fixed by the Regulations.

Article 20

The details of the application of this Agreement shall be determined by Regulations the provisions of which may, at any time, be amended with the common consent of the Offices of the contracting countries.

Article 21

The provisions of this Agreement offer only a minimum of protection; they shall not preclude the claiming of the application of wider provisions that may be enacted by the domestic legislation of a contracting country, nor shall they prejudice the application of the provisions of the Berne Convention, as revised in 1928, relating to the protection of artistic works and works of art applied to industry.

Article 22

(1) Countries members of the Union which are not party to this Agreement may accede thereto at their request and in the manner prescribed by Articles 16 and 16bis of the General Convention.

(2) Notification of accession shall, of itself, ensure, upon the territory of the acceding country, the benefits of the foregoing provisions to industrial designs which, at the time of accession, are the subject of international deposit.

(3) However, any country may, in acceding to this Agreement, declare that application of this Act shall be limited to designs deposited from the date on which its accession becomes effective.

(4) In the case of denunciation of this Agreement, Article 17bis of the General Convention shall apply. International designs deposited up to the date on which denunciation becomes effective shall continue, throughout the period of international protection, to enjoy in the denouncing country, as well as in all other countries of the restricted Union, the same protection as if they had been deposited direct in such Countries.

Article 23

(1) This Agreement shall be ratified and ratifications shall be deposited at London not later than July 1, 1938.

(2) It shall enter into force, between the countries which have ratified it, one month after that date, and shall have the same force and duration as the General Convention.

(3) This Act shall, as regards the relations between the countries which have ratified it, replace the Hague Agreement of 1925. However, the latter shall remain in force as regards the relations with countries which have not ratified this Act.

APPENDIX V

Hague Act of November 28, 1960

Table of Contents

Article 1

(1) The contracting States constitute a Special Union for the international deposit of industrial designs.

(2) Only States members of the International Union for the Protection of Industrial Property may become party to this Agreement.

Article 2

For the purposes of this Agreement:

"1925 Agreement" shall mean the Hague Agreement concerning the International Deposit of Industrial Designs of November 6, 1925;

"1934 Agreement" shall mean the Hague Agreement concerning the International Deposit of Industrial Designs of November 6, 1925, as revised at London on June 2, 1934;

"this Agreement" or "the present Agreement" shall mean the Hague Agreement concerning the International Deposit of Industrial Designs as established by the present Act;

"Regulations" shall mean the Regulations for carrying out this Agreement;

"International Bureau" shall mean the International Bureau of Intellectual Property;

"international deposit" shall mean a deposit made at the International Bureau;

"national deposit" shall mean a deposit made at the national Office of a contracting State;

"multiple deposit" shall mean a deposit including several designs;

"State of origin of an international deposit" shall mean the contracting State in which the applicant has a real and effective industrial or commercial establishment or, if the applicant has such establishments in several contracting States, the contracting State which he has indicated in his application; if the applicant has no such establishment in any contracting State, the contracting State in which he has his domicile; if he has no domicile in a contracting State, the contracting State of which he is a national;

"State having a novelty examination" shall mean a contracting State the domestic law of which provides for a system which involves a preliminary *ex officio* search and examination by its national Office as to the novelty of each deposited design.

Article 3

Nationals of contracting States and persons who, without being nationals of any contracting State, are domiciled or have a real and effective industrial or commercial establishment in the territory of a contracting State may deposit designs at the International Bureau.

Article 4

(1) International deposit may be made at the International Bureau:

1. direct, or
2. through the intermediary of the national Office of a contracting State if the law of that State so permits.

(2) The domestic law of any contracting State may require that international deposits of which it is deemed to be the State of origin shall be made through its national Office. Non-compliance with this requirement shall not prejudice the effects of the international deposit in the other contracting States.

Article 5

(1) The international deposit shall consist of an application and one or more photographs or other graphic representations of the design, and shall involve payment of the fees prescribed by the Regulations.

(2) The application shall contain:

1. a list of the contracting States in which the applicant requests that the international deposit shall have effect;
2. the designation of the article or articles in which it is intended to incorporate the design;
3. if the applicant wishes to claim the priority provided for in Article 9, an indication of the date, the State, and the number of the deposit giving rise to the right of priority;
4. such other particulars as the Regulations may prescribe.

(3) *(a)* In addition, the application may contain:

1. a short description of characteristic features of the design;
2. a declaration as to who is the true creator of the design;
3. a request for deferment of publication as provided in Article 6(4).

(b) The application may be accompanied also by samples or models of the article or articles incorporating the design.

(4) A multiple deposit may include several designs intended to be incorporated in articles included in the same class of the International Design Classification referred to in Article 21(2), item 4.

Article 6

(1) The International Bureau shall maintain the International Design Register and shall register international deposits therein.

(2) The international deposit shall be deemed to have been made on the date on which the International Bureau received the application in due form, the fees payable with the application, and the photograph or photographs or other graphic representations of the design, or, if the International Bureau received them on different dates, on the last of these dates. The registration shall bear the same date.

(3) *(a)* For each international deposit, the International Bureau shall publish in a periodical bulletin:

1. reproductions in black and white or, at the request of the applicant, in color of the deposited photographs or other graphic representations;
2. the date of the international deposit;
3. the particulars prescribed by the Regulations.

(b) The International Bureau shall send the periodical bulletin to the national Offices as soon as possible.

(4) *(a)* The publication referred to in paragraph (3)*(a)* shall, at the request of the applicant, be deferred for such period as he may request. The said period may not exceed twelve months from the date of the international deposit. However, if priority is claimed, the starting date of such period shall be the priority date.

(b) At any time during the period referred to in subparagraph *(a)*, the applicant may request immediate publication or may withdraw his deposit. Withdrawal of the deposit may be limited to one or a few only of the contracting States and, in the case of a multiple deposit, to some only of the designs included therein.

(c) If the applicant fails to pay within the proper time the fees payable before the expiration of the period referred to in subparagraph *(a)*, the International Bureau shall cancel the deposit and shall not effect the publication referred to in paragraph (3)*(a)*.

(d) Until the expiration of the period referred to in subparagraph *(a)*, the International Bureau shall keep in confidence the registration of deposits made subject to deferred publication, and the public shall have no access to any documents or articles concerning such deposits. These provisions shall apply without limitation as to time if the applicant has withdrawn his deposit before the expiration of the said period.

(5) Except as provided in paragraph (4), the Register and all documents and articles filed with the International Bureau shall be open to inspection by the public.

Article 7

(1) *(a)* A deposit registered at the International Bureau shall have the same effect in each of the contracting States designated by the applicant in his application as if all the formalities required by the domestic law for the grant of protection had been complied with by the applicant and as if all administrative acts required to that end had been accomplished by the Office of such State.

(b) Subject to the provisions of Article 11, the protection of designs the deposit of which has been registered at the International Bureau is governed in each contracting State by those provisions of the domestic law which are applicable in that State to designs for which protection has been claimed on the basis of a national deposit and in respect of which all formalities and administrative acts have been complied with and accomplished.

(2) An international deposit shall have no effect in the State of origin if the laws of that State so provide.

Article 8

(1) Notwithstanding the provisions of Article 7, the national Office of a contracting State whose domestic law provides that the national Office may, on the basis of an administrative *ex officio* examination or pursuant to an opposition by a third party, refuse protection shall, in case of refusal, notify the International Bureau within six months that the design does not meet the requirements of its domestic law other than the formalities and administrative acts referred to in Article 7(1). If no such refusal is notified within a period of six months the international deposit shall become effective in that State as from the date of that deposit. However, in a contracting State having a novelty examination, the international deposit, while

retaining its priority, shall, if no refusal is notified within a period of six months, become effective from the expiration of the said period unless the domestic law provides for an earlier date for deposits made with its national Office.

(2) The period of six months referred to in paragraph (1) shall be computed from the date on which the national Office receives the issue of the periodical bulletin in which the registration of the international deposit has been published. The national Office shall communicate that date to any person so requesting.

(3) The applicant shall have the same remedies against the refusal of the national Office referred to in paragraph (1) as if he had deposited his design in that Office; in any case, the refusal shall be subject to a request for re-examination or appeal. Notification of such refusal shall indicate:

1. the reasons for which it has been found that the design does not meet the requirements of the domestic law;
2. the date referred to in paragraph (2);
3. the time allowed for a request for re-examination or appeal;
4. the authority to which such request or appeal may be addressed.

(4) *(a)* The national Office of a contracting State whose domestic law contains provisions of the kind referred to in paragraph (1) requiring a declaration as to who is the true creator of the design or a description of the design may provide that, upon request and within a period of not less than sixty days from the dispatch of such a request by the said Office, the applicant shall file in the language of the application filed with the International Bureau:

1. a declaration as to who is the true creator of the design;
2. a short description emphasizing the essential characteristic features of the design as shown by the photographs or other graphic representations.

(b) No fees shall be charged by a national Office in connection with the filing of such declarations or descriptions, or for their possible publication by that national Office.

(5) *(a)* Any contracting State whose domestic law contains provisions of the kind referred to in paragraph (1) shall notify the International Bureau accordingly.

(b) If, under its legislation, a contracting State has several systems for the protection of designs one of which provides for novelty examination, the provisions of this Agreement concerning States having a novelty examination shall apply only to the said system.

Article 9

If the international deposit of a design is made within six months of the first deposit of the same design in a State member of the International Union for the Protection of Industrial Property, and if priority is claimed for the international deposit, the priority date shall be that of the first deposit.

Article 10

(1) An international deposit may be renewed every five years by payment only, during the last year of each period of five years, of the renewal fees prescribed by the Regulations.

(2) Subject to the payment of a surcharge fixed by the Regulations, a period of grace of six months shall be granted for renewal of the international deposit.

(3) At the time of paying the renewal fees, the international deposit number must be indicated and also, if renewal is not to be effected for all the contracting States for which the deposit is about to expire, those of the contracting States for which the renewal is to be effected.

(4) Renewal may be limited to some only of the designs included in a multiple deposit.

(5) The International Bureau shall record and publish renewals.

Article 11

(1) *(a)* The term of protection granted by a contracting State to designs which have been the subject of an international deposit shall not be less than:

1. ten years from the date of the international deposit if the deposit has been renewed;
2. five years from the date of the international deposit in the absence of renewal.

(b) However, if, under the provisions of the domestic law of a contracting State having a novelty examination, protection commences at a date later than that of the international deposit, the minimum terms provided for in subparagraph *(a)* shall be computed from the date at which protection commences in that State. The fact that the international deposit is not renewed or is renewed only once shall in no way affect the minimum terms of protection thus defined.

(2) If the domestic law of a contracting State provides, in respect of designs which have been the subject of a national deposit, for protection whose duration, with or

without renewal, is longer than ten years, protection of the same duration shall, on the basis of the international deposit and its renewals, be granted in that State to designs which have been the subject of an international deposit.

(3) A contracting State may, under its domestic law, limit the term of protection of designs which have been the subject of an international deposit to the terms provided for in paragraph (1).

(4) Subject to the provisions of paragraph (1)*(b)*, protection in a contracting State shall terminate at the date of expiration of the international deposit, unless the domestic law of that State provides that protection shall continue after the date of expiration of the international deposit.

Article 12

(1) The International Bureau shall record and publish changes affecting ownership of a design which is the subject of an international deposit in force. It is understood that transfer of ownership may be limited to the rights arising from the international deposit in one or a few only of the contracting States and, in the case of a multiple deposit, to some only of the designs included therein.

(2) The recording referred to in paragraph (1) shall have the same effect as if it had been made in the national Offices of the contracting States.

Article 13

(1) The owner of an international deposit may, by means of a declaration addressed to the International Bureau, renounce his rights in respect of all or some only of the contracting States and, in the case of a multiple deposit, in respect of some only of the designs included therein.

(2) The International Bureau shall record and publish such declaration.

Article 14

(1) No contracting State may, as a condition of recognition of the right to protection, require that the article incorporating the design bear a sign or notice concerning the deposit of the design.

(2) If the domestic law of a contracting State provides for a notice on the article for any other purpose, such State shall regard such requirement as satisfied if all the articles offered to the public with the authorization of the owner of the rights in the design, or the tags attached to such articles, bear the international design notice.

(3) The international design notice shall consist of the symbol Ⓓ accompanied by:

1. the year of the international deposit and the name, or the usual abbreviation of the name, of the depositor, or
2. the number of the international deposit.

(4) The mere appearance of the international design notice on the article or the tags shall in no case be interpreted as implying a waiver of protection by virtue of copyright or on any other grounds, whenever, in the absence of such notice, such protection may be claimed.

Article 15

(1) The fees prescribed by the Regulations shall consist of:

1. fees for the International Bureau;
2. fees for the contracting States designated by the applicant, namely:
   (a) a fee for each contracting State;
   (b) a fee for each contracting State having a novelty examination and requiring the payment of a fee for such examination.

(2) Any fees paid in respect of one and the same deposit for a contracting State under paragraph (1), item 2*(a)*, shall be deducted from the amount of the fee referred to in paragraph (1), item 2*(b)*, if the latter fee becomes payable for the same State.

Article 16

(1) The fees for contracting States referred to in Article 15(1), item 2, shall be collected by the International Bureau and paid over annually to the contracting States designated by the applicant.

(2) *(a)* Any contracting State may notify the International Bureau that it waives its right to the supplementary fees referred to in Article 15(1), item 2*(a)*, in respect of international deposits of which any other contracting State making a similar waiver is deemed to be the State of origin.

(b) Such State may make a similar waiver in respect of international deposits of which it is itself deemed to be the State of origin.

Article 17

The Regulations shall govern the details concerning the implementation of this Agreement and in particular:

1. the languages and the number of copies in which the application for deposit must be filed, and the data to be supplied in the application;

2. the amounts and the dates and method of payment of the fees for the International Bureau and for the States, including the limits imposed on the fee for contracting States having a novelty examination;
3. the number, size, and other characteristics, of the photographs or other graphic representations of each design deposited;
4. the length of the description of characteristic features of the design;
5. the limits within which and conditions under which samples or models of the articles incorporating the design may accompany the application;
6. the number of designs that may be included in a multiple deposit and other conditions governing multiple deposits;
7. all matters relating to the publication and distribution of the periodical bulletin referred to in Article 6(3)*(a)*, including the number of copies of the bulletin which shall be given free of charge to the national Offices and the number of copies which may be sold at a reduced price to such Offices;
8. the procedure for notification by contracting States of any refusal provided for under Article 8(1), and the procedure for communication and publication of such refusals by the International Bureau;
9. the conditions for recording and publication by the International Bureau of the changes affecting the ownership of a design referred to in Article 12(1), and for the renunciations referred to in Article 13;
10. the disposal of documents and articles concerning deposits for which the possibility of renewal has ceased to exist.

Article 18

The provisions of this Agreement shall not preclude the making of a claim to the benefit of any greater protection which may be granted by domestic legislation in a contracting State, nor shall they affect in any way the protection accorded to works of art and works of applied art by international copyright treaties and conventions.

Articles 19 to 22

[Repealed by Article 7(2) of the Complementary Act of Stockholm (1967)]

Article 23

(1) This Agreement shall remain open for signature until December 31, 1961.

(2) It shall be ratified and the instruments of ratification shall be deposited with the Government of the Netherlands.

Article 24

(1) States members of the International Union for the Protection of Industrial Property which have not signed this Agreement may accede thereto.

(2) Such accessions shall be notified through diplomatic channels to the Director General,* and by the latter to the Governments of all contracting States.

Article 25

(1) Each contracting State undertakes to provide for the protection of industrial designs and to adopt, in accordance with its constitution, the measures necessary to ensure the application of this Agreement.

(2) At the time a contracting State deposits its instrument of ratification or accession, it must be in a position under its domestic law to give effect to the provisions of this Agreement.

Article 26

(1) This Agreement shall enter into force one month after the date on which the Director General has dispatched a notification to the contracting States of the deposit of ten instruments of ratification or accession, at least four of which are those of States which, at the date of the present Agreement, are not party either to the 1925 Agreement or to the 1934 Agreement.

(2) Thereafter, the deposit of instruments of ratification and accession shall be notified to the contracting States by the Director General. Such ratifications and accessions shall become effective one month after the date of the dispatch of such notification unless, in the case of accession, a later date is indicated in the instrument of accession.

Article 27

Any contracting State may at any time notify the Director General that this Agreement shall also apply to all or part of those territories for the external relations of which it is responsible. Thereupon, the Director General shall communicate such

---

* "Director General" is defined by Article 1 of the Complementary Act of Stockholm (1967) as the Director General of the World Intellectual Property Organization

notification to the contracting States and the Agreement shall apply also to the said territories one month after the dispatch of the communication by the Director General to the contracting States unless a later date is indicated in the notification.

Article 28

(1) Any contracting State may, by notification addressed to the Director General, denounce this Agreement in its own name and on behalf of all or part of the territories designated in the notification under Article 27. Such notification shall take effect one year after its receipt by the Director General.

(2) Denunciation shall not relieve any contracting State of its obligations under this Agreement in respect of designs deposited at the International Bureau prior to the date on which the denunciation takes effect.

Article 29

(1) This Agreement shall be submitted to revision with a view to the introduction of amendments designed to improve the protection resulting from the international deposit of designs.

(2) Revision conferences shall be called at the request of not less than one-half of the contracting States.

Article 30

(1) Two or more contracting States may at any time notify the Director General that, subject to the conditions indicated in the notification:

1. a common Office shall be substituted for the national Office of each of them;
2. they shall be deemed to be a single State for the purposes of the application of Articles 2 to 17 of this Agreement.

(2) Such notification shall not take effect until six months after the date of dispatch of the communication thereof by the Director General to the other contracting States.

Article 31

(1) This Agreement alone shall be applicable as regards the mutual relations of States party to both the present Agreement and the 1925 Agreement or the 1934 Agreement. However, such States shall, in their mutual relations, apply the 1925 Agreement or the 1934 Agreement, as the case may be, to designs deposited at the

International Bureau prior to the date on which the present Agreement becomes applicable as regards their mutual relations.

(2) *(a)* Any State party to both the present Agreement and the 1925 Agreement shall continue to apply the 1925 Agreement in its relations with States party only to the 1925 Agreement, unless the said State has denounced the 1925 Agreement.

(b) Any State party to both the present Agreement and the 1934 Agreement shall continue to apply the 1934 Agreement in its relations with States party only to the 1934 Agreement, unless the said State has denounced the 1934 Agreement.

(3) States party to the present Agreement only shall not be bound to States which, without being party to the present Agreement, are party to the 1925 Agreement or the 1934 Agreement.

Article 32

(1) Signature and ratification of, or accession to, the present Agreement by a State party, at the date of this Agreement, to the 1925 Agreement or the 1934 Agreement shall be deemed to include signature and ratification of, or accession to, the Protocol annexed to the present Agreement, unless such State makes an express declaration to the contrary at the time of signing or depositing its instrument of accession.

(2) Any contracting State having made the declaration referred to in paragraph (1), or any other contracting State not party to the 1925 Agreement or the 1934 Agreement, may sign or accede to the Protocol annexed to this Agreement. At the time of signing or depositing its instrument of accession, it may declare that it does not consider itself bound by the provisions of paragraphs (2)*(a)* or (2)*(b)* of the Protocol; in such case, the other States party to the Protocol shall be under no obligation to apply, in their relations with that State, the provisions mentioned in such declaration. The provisions of Articles 23 to 28 inclusive shall apply by analogy.

Article 33

This Act shall be signed in a single copy which shall be deposited in the archives of the Government of the Netherlands. A certified copy shall be transmitted by the latter to the Government of each State which has signed or acceded to this Agreement.

*Protocol**

States party to this Protocol have agreed as follows:

(1) The provisions of this Protocol shall apply to designs which have been the subject of an international deposit and of which one of the States party to this Protocol is deemed to be the State of origin.

(2) In respect of designs referred to in paragraph (1), above:

(a) the term of protection granted by States party to this Protocol to the designs referred to in paragraph (1) shall not be less than fifteen years from the date provided for in paragraphs (1)*(a)* or (1)*(b)*, as the case may be, of Article 11;

(b) the appearance of a notice on the articles incorporating the designs or on the tags attached thereto shall in no case be required by the States party to this Protocol, either for the exercise in their territories of rights arising from the international deposit, or for any other purpose.

---

* This Protocol is not yet in force

APPENDIX VI

Additional Act of Monaco of November 18, 1961

Table of Contents

Article 1

(1) Over and above the fees established in Article 15 of the Hague Agreement as revised at London, the following additional fees shall be payable in respect of the transactions hereinafter specified, that is to say:

1. for the deposit of a single design and in respect of the first period of five years: 20 Swiss francs;
2. for the deposit of a single design, upon expiration of the first period and in respect of the duration of the second period of ten years: 40 Swiss francs;
3. for a multiple deposit and in respect of the first period of five years: 50 Swiss francs;
4. for a multiple deposit, upon expiration of the first period and in respect of the duration of the second period of ten years: 200 Swiss francs.

(2) If the fees prescribed in Article 15.2 and Article 15.4 of the Hague Agreement as revised at London have been paid after the date of this Act but before its entry into force – the latter being determined for each State in accordance with the provisions of paragraphs 7(2) and 7(3) where the first period of protection expires after such entry into force, the person making the deposit must pay the additional prolongation fee specified in items (1)2 and (1)4 of this Article. Upon entry into force of this Act, the International Bureau shall advise the depositors concerned

that they must pay the additional fee within a period of six months from the receipt of such notice. If payment is not effected within such period the prolongation shall be deemed to be null and the reference thereto shall be deleted from the Register. In such case, the fee for prolongation already paid shall be refunded.

Article 2

Additional fees of 20 Swiss francs or 10 Swiss francs shall likewise be payable in respect of every other transaction provided for by the Hague Agreement as revised at London and for which the Regulations of the said Agreement prescribe a fee of 5 Swiss francs or 2.50 Swiss francs.

Article 3

(1) The fees prescribed in Articles 1 and 2 of this Act may be modified on the proposal of the International Bureau or of the Swiss Government, in accordance with the procedure hereinafter defined.

(2) Such proposals shall be communicated to the Offices of States party to this Act, which shall communicate their views to the International Bureau within a period of six months. If, after that period, a modification of a fee is adopted by a majority of the said Offices without giving rise to any opposition, such modification shall enter into force on the first day of the month following the date of dispatch of the notification thereof by the International Bureau to the aforesaid Offices.

Article 4

(1) There shall be established from the excess receipts derived from the application of the additional fees a reserve fund not exceeding 50,000 Swiss francs.

(2) When the reserve fund has reached this amount, any further excess receipts shall be distributed among the States party to this Act in proportion to the number of designs deposited by their nationals, or by the other persons referred to in Article 1 of the Hague Agreement as revised at London.

Article 5

For such time as all countries members of the Union created by the Hague Agreement as revised at London are not party to this Act or to the Hague Agreement of November 28, 1960, the International Bureau shall draw up separate accounts for countries which are party to this Act and for those which are party only to the Hague Agreement as revised at London.

Article 6

(1) This Act shall remain open for signature until March 31, 1962.

(2) States party to the Hague Agreement as revised at London which have not signed this Act may accede thereto. In such cases, the provisions of Articles 16 and 16bis of the Paris Convention for the Protection of Industrial Property shall be applicable.

Article 7

(1) This Act shall be ratified and the instruments of ratification shall be deposited with the Government of the Principality of Monaco. Such deposits shall be notified by the latter Government to the Government of the Swiss Confederation, which shall notify them to the contracting States.

(2) This Act shall come into force at the expiration of a period of one month from the date of dispatch by the Government of the Swiss Confederation to the contracting States of the notification of the deposit of the second instrument of ratification.

(3) As regards States which deposit their instruments of ratification subsequently to the deposit of the second such instrument of ratification as is mentioned in the preceding paragraph, this Act shall enter into force upon expiration of a period of one month from the date of dispatch by the Government of the Swiss Confederation to the contracting States of the notification of the deposit of the instrument of ratification concerned.

Article 8

This Act shall be signed in a single copy and shall be deposited in the archives of the Government of the Principality of Monaco. A certified copy thereof shall be sent by the latter to each of the Governments of the countries of the Hague Union.

APPENDIX VII

Complementary Act of Stockholm of July 14, 1967

Table of Contents

Article 1
[Definitions]

For the purposes of this Complementary Act:

"1934 Act" shall mean the Act signed at London on June 2, 1934, of the Hague Agreement concerning the International Deposit of Industrial Designs;

"1960 Act" shall mean the Act signed at The Hague on November 28, 1960, of the Hague Agreement concerning the International Deposit of Industrial Designs;

"1961 Additional Act" shall mean the Act signed at Monaco on November 18, 1961, additional to the 1934 Act;

"Organization" shall mean the World Intellectual Property Organization;

"International Bureau" shall mean the International Bureau of Intellectual Property;

"Director General" shall mean the Director General of the Organization;

"Special Union" shall mean the Hague Union established by the Hague Agreement of November 6, 1925, concerning the International Deposit of Industrial

Designs, and maintained by the 1934 and 1960 Acts, by the 1961 Additional Act, and by this Complementary Act.

Article 2
[Assembly]

(1) (a) The Special Union shall have an Assembly consisting of those countries which have ratified or acceded to this Complementary Act.

(b) The Government of each country shall be represented by one delegate, who may be assisted by alternate delegates, advisors, and experts.

(c) The expenses of each delegation shall be borne by the Government which has appointed it.

(2) (a) The Assembly shall:

(i) deal with all matters concerning the maintenance and development of the Special Union and the implementation of this Agreement;

(ii) give directions to the International Bureau concerning the preparation for conferences of revision, due account being taken of any comments made by those countries of the Special Union which have not ratified or acceded to this Complementary Act;

(iii) modify the Regulations, including the fixation of the amounts of the fees relating to the international deposit of industrial designs;

(iv) review and approve the reports and activities of the Director General concerning the Special Union, and give him all necessary instructions concerning matters within the competence of the Special Union;

(v) determine the program and adopt the biennial budget of the Special Union, and approve its final accounts;

(vi) adopt the financial regulations of the Special Union;

(vii) establish such committees of experts and working groups as it may deem necessary to achieve the objectives of the Special Union;

(viii) determine which countries not members of the Special Union and which intergovernmental and international non-governmental organizations shall be admitted to its meetings as observers;

(ix) adopt amendments to Articles 2 to 5;

(x) take any other appropriate action designed to further the objectives of the Special Union;

(xi) perform such other functions as are appropriate under this Complementary Act.

(b) With respect to matters which are of interest also to other Unions administered by the Organization, the Assembly shall make its decisions after having heard the advice of the Coordination Committee of the Organization.

(3) (a) Each country member of the Assembly shall have one vote.

(b) One-half of the countries members of the Assembly shall constitute a quorum.

(c) Notwithstanding the provisions of subparagraph (b), if, in any session, the number of countries represented is less than one-half but equal to or more than one-third of the countries members of the Assembly, the Assembly may make decisions but, with the exception of decisions concerning its own procedure, all such decisions shall take effect only if the conditions set forth hereinafter are fulfilled. The International Bureau shall communicate the said decisions to the countries members of the Assembly which were not represented and shall invite them to express in writing their vote or abstention within a period of three months from the date of the communication. If, at the expiration of this period, the number of countries having thus expressed their vote or abstention attains the number of countries which was lacking for attaining the quorum in the session itself, such decisions shall take effect provided that at the same time the required majority still obtains.

(d) Subject to the provisions of Article 5(2), the decisions of the Assembly shall require two-thirds of the votes cast.

(e) Abstentions shall not be considered as votes.

(f) A delegate may represent, and vote in the name of, one country only.

(g) Countries of the Special Union not members of the Assembly shall be admitted to the meetings of the latter as observers.

(4) (a) The Assembly shall meet once in every second calendar year in ordinary session upon convocation by the Director General and, in the absence of exceptionnal circumstances, during the same period and at the same place as the General Assembly of the Organization.

(b) The Assembly shall meet in extraordinary session upon convocation by the Director General, at the request of one-fourth of the countries members of the Assembly.

(c) The agenda of each session shall be prepared by the Director General.

(5) The Assembly shall adopt its own rules of procedure.

Article 3
[International Bureau]

(1) (a) International deposit of industrial designs and related duties, as well as all other administrative tasks concerning the Special Union, shall be performed by the International Bureau.

(b) In particular, the International Bureau shall prepare the meetings and provide the secretariat of the Assembly and of such committees of experts and working groups as may have been established by the Assembly.

(c) The Director General shall be the chief executive of the Special Union and shall represent the Special Union.

(2) The Director General and any staff member designated by him shall participate, without the right to vote, in all meetings of the Assembly and of such committees of experts or working groups as may have been established by the Assembly. The Director General, or a staff member designated by him, shall be ex officio secretary of those bodies.

(3) (a) The International Bureau shall, in accordance with the directions of the Assembly, make the preparations for the conferences of revision of the provisions of the Agreement.

(b) The International Bureau may consult with intergovernmental and international non-governmental organizations concerning preparations for conferences of revision.

(c) The Director General and persons designated by him shall take part, without the right to vote, in the discussions at those conferences.

(4) The International Bureau shall carry out any other tasks assigned to it.

Article 4
[Finances]

(1) (a) The Special Union shall have a budget.

(b) The budget of the Special Union shall include the income and expenses proper to the Special Union, its contribution to the budget of expenses common to the Unions, and, where applicable, the sum made available to the budget of the Conference of the Organization.

(c) Expenses not attributable exclusively to the Special Union but also to one or more other Unions administered by the Organization shall be deemed to be expenses common to the Unions. The share of the Special Union in such common expenses shall be in proportion to the interest the Special Union has in them.

(2) The budget of the Special Union shall be established with due regard to the requirements of coordination with the budgets of the other Unions administered by the Organization.

(3) The budget of the Special Union shall be financed from the following sources:

(i) international deposit fees and other fees and charges due for other services rendered by the International Bureau in relation to the Special Union;

(ii) sale of, or royalties on, the publications of the International Bureau concerning the Special Union;

(iii) gifts, bequests, and subventions;

(iv) rents, interests, and other miscellaneous income.

(4) (a) The amounts of the fees referred to in paragraph (3)(i) shall be fixed by the Assembly on the proposal of the Director General.

(b) The amounts of such fees shall be so fixed that the revenues of the Special Union from fees and other sources shall be at least sufficient to cover the expenses of the International Bureau concerning the Special Union.

(c) If the budget is not adopted before the beginning of a new financial period, it shall be at the same level as the budget of the previous year, as provided in the financial regulations.

(5) Subject to the provisions of paragraph (4)(a), the amount of the fees and charges due for other services rendered by the International Bureau in relation to the Special Union shall be established, and shall be reported to the Assembly, by the Director General.

(6) (a) The Special Union shall have a working capital fund which shall be constituted by the excess receipts and, if such excess does not suffice, by a single payment made by each country of the Special Union. If the fund becomes insufficient, the Assembly shall decide to increase it.

(b) The amount of the initial payment of each country to the said fund or of its participation in the increase thereof shall be a proportion of the contribution of that country as a member of the Paris Union for the Protection of Industrial Property to the budget of the said Union for the year in which the fund is established or the decision to increase it is made.

(c) The proportion and the terms of payment shall be fixed by the Assembly on the proposal of the Director General and after it has heard the advice of the Coordination Committee of the Organization.

(7) (a) In the headquarters agreement concluded with the country on the territory of which the Organization has its headquarters, it shall be provided that, whenever the working capital fund is insufficient, such country shall grant advances. The amount of those advances and the conditions on which they are granted shall be the subject of separate agreements, in each case, between such country and the Organization.

(b) The country referred to in subparagraph (a) and the Organization shall each have the right to denounce the obligation to grant advances, by written notification. Denunciation shall take effect three years after the end of the year in which it has been notified.

(8) The auditing of the accounts shall be effected by one or more of the countries of the Special Union or by external auditors, as provided in the financial regulations. They shall be designated, with their agreement, by the Assembly.

Article 5
[Amendment of Articles 2 to 5]

(1) Proposals for the amendment of this Complementary Act may be initiated by any country member of the Assembly, or by the Director General. Such proposals shall be communicated by the Director General to the member countries of the Assembly at least six months in advance of their consideration by the Assembly.

(2) Amendments referred to in paragraph (1) shall be adopted by the Assembly. Adoption shall require three-fourths of the votes cast, provided that any amendment to Article 2 and to the present paragraph, shall require four-fifths of the votes cast.

(3) Any amendment referred to in paragraph (1) shall enter into force one month after written notifications of acceptance, effected in accordance with their respective constitutional processes, have been received by the Director General from three-fourths of the countries members of the Assembly at the time it adopted the amendment. Any amendment thus accepted shall bind all the countries which are members of the Assembly at the time the amendment enters into force, or which become members thereof at a subsequent date.

Article 6
[Amendment of the 1934 Act and the 1961 Additional Act]

(1) (a) References in the 1934 Act to "the International Bureau of Industrial Property at Berne," to "the Berne International Bureau," or to "the International Bureau," shall be construed as references to the International Bureau as defined in Article 1 of this Complementary Act.

(b) Article 15 of the 1934 Act is repealed.

(c) Any amendment of the Regulations referred to in Article 20 of the 1934 Act shall be effected in accordance with the procedure prescribed under Article 2(2)(a)(iii) and Article 2(3)(d).

(d) In Articles 21 of the 1934 Act, for the words "revised in 1928" there shall be substituted the words "for the Protection of Literary and Artistic Works."

(e) References in Article 22 of the 1934 Act to Articles 16, 16bis, and 17bis, of the "General Convention" shall be construed as references to those provisions of the Stockholm Act of the Paris Convention for the Protection of Industrial Property which, in the said Stockholm Act, correspond to Articles 16, 16bis, and 17bis, of the earlier Acts of the Paris Convention.

(2) (a) Any modification of the fees referred to in Article 3 of the 1961 Additional Act shall be effected in accordance with the procedure prescribed under Article 2(2)(a)(iii) and 2(3)(d).

(b) Article 4(1) of the 1961 Additional Act, and the words "When the reserve fund has reached this amount" in Article 4(2) are repealed.

(c) References in Article 6(2) of the 1961 Additional Act to Articles 16 and 16bis of the Paris Convention for the Protection of Industrial Property shall be construed as references to those provisions of the Stockholm Act of the said Convention which, in the Stockholm Act, correspond to Articles 16 and 16bis of the earlier Acts of the Paris Convention.

(d) References in paragraph 7(1) and paragraph 7(3) of the 1961 Additional Act to the Government of the Swiss Confederation shall be construed as references to the Director General.

Article 7
[Amendment of the 1960 Act]

(1) References in the 1960 Act to "the Bureau of the International Union for the Protection of Industrial Property" or to "the International Bureau" shall be

construed as references to the International Bureau as defined in Article 1 of this Complementary Act.

(2) Articles 19, 20, 21, and 22, of the 1960 Act are repealed.

(3) References in the 1960 Act to the Government of the Swiss Confederation shall be construed as references to the Director General.

(4) In Article 29 of the 1960 Act, the words "periodical" (paragraph 29(1)) and "of the International Design Committee or" (paragraph 29(2)) are deleted.

Article 8
[Ratification of, and Accession to, the Complementary Act]

(1) (a) Countries which, before January 13, 1968, have ratified the 1934 Act or the 1960 Act, and countries which have acceded to at least one of those Acts, may sign this Complementary Act and ratify it, or may accede to it.

(b) Ratification of, or accession to, this Complementary Act by a country which is bound by the 1934 Act without being bound also by the 1961 Additional Act shall automatically entail ratification of, or accession to, the 1961 Additional Act.

(2) Instruments of ratification and accession shall be deposited with the Director General.

Article 9
[Entry Into Force of the Complementary Act]

(1) With respect to the first five countries which have deposited their instruments of ratification or accession, this Complementary Act shall enter into force three months after the deposit of the fifth such instrument of ratification or accession.

(2) With respect to any other country, this Complementary Act shall enter into force three months after the date on which its ratification or accession has been notified by the Director General, unless a subsequent date has been indicated in the instrument of ratification or accession. In the latter case, this Complementary Act shall enter into force with respect to that country on the date thus indicated.

Article 10
[Automatic Acceptance of Certain Provisions by Certain Countries]

(1) Subject to the provisions of Article 8 and the following paragraph, any country which has not ratified or acceded to the 1934 Act shall become bound by the 1961 Additional Act and by Articles 1 to 6 of this Complementary Act from the date on which its accession to the 1934 Act enters into force, provided that, if on the said date this Complementary Act has not yet entered into force pursuant to Article 9(1),

then, such country shall become bound by the said Articles of this Complementary Act only from the date of entry into force of the Complementary Act pursuant to Article 9(1).

(2) Subject to the provisions of Article 8 and the foregoing paragraph, any country which has not ratified or acceded to the 1960 Act shall become bound by Articles 1 to 7 of this Complementary Act from the date on which its ratification of, or accession to, the 1960 Act enters into force, provided that, if on the said date this Complementary Act has not yet entered into force pursuant to Article 9(1), then, such country shall become bound by the said Articles of this Complementary Act only from the date of entry into force of the Complementary Act pursuant to Article 9(1).

Article 11
[Signature, etc., of the Complementary Act]

(1) (a) This Complementary Act shall be signed in a single copy in the French language and shall be deposited with the Government of Sweden.

(b) Official texts shall be established by the Director General, after consultation with the interested Governments, in such other languages as the Assembly may designate.

(2) This Complementary Act shall remain open for signature at Stockholm until January 13, 1968.

(3) The Director General shall transmit two copies, certified by the Government of Sweden, of the signed text of this Complementary Act to the Governments of all countries of the Special Union and, on request, to the Government of any other country.

(4) The Director General shall register this Complementary Act with the Secretariat of the United Nations.

(5) The Director General shall notify the Governments of all countries of the Special Union of signatures, deposits of instruments of ratification or accession, entry into force, and all other relevant notifications.

Article 12
[Transitional Provision]

Until the first Director General assumes office, references in this Complementary Act to the International Bureau of the Organization or to the Director General shall be construed as references to the Bureau of the Union established by the Paris Convention for the Protection of Industrial Property or its Director, respectively.

APPENDIX VIII

International Registration Application Form DM/I(E)-IX.04

DM/1(E)

HAGUE AGREEMENT

CONCERNING THE INTERNATIONAL REGISTRATION OF INDUSTRIAL DESIGNS

APPLICATION FOR INTERNATIONAL REGISTRATION

GOVERNED EXCLUSIVELY OR PARTLY BY THE 1999 ACT AND/OR THE 1960 ACT

IMPORTANT

1. The international application may contain **several industrial designs**. However, all the industrial designs and/or products in relation to which such designs are to be used must belong to the **same class** of the International (Locarno) Classification.
2. Photographs and other graphic representations of the industrial designs should be pasted or printed directly onto a separate sheet of A4 paper, white and opaque, in compliance with the instructions made available on document DM/1.inf (paragraphs 92 to 111).
3. A **Fee Calculator** is available on the WIPO Website under the headings *"Activities and Services/International Designs/Fees/Automated Fee Calculator"*.
4. This form must be typed and not be completed by handwriting. An editable PDF format is also available on the WIPO Website under the headings *"Activities and Services/International Designs/Fees/Forms"*.

This cover page must not be sent to the International Bureau.

World Intellectual Property Organization
34, chemin des Colombettes, P.O. Box 18,
1211 Geneva 20, Switzerland
Tel.: (41-22) 338 9111
Fax (International Trademark Registry): (41-22) 740 1429
e-mail: intreg.mail@wipo.int – Internet: *http://www.wipo.int*

DM/1 (E) – IX.04

DM/1(E)

APPLICATION FOR INTERNATIONAL REGISTRATION GOVERNED EXCLUSIVELY OR PARTLY BY THE 1999 ACT AND/OR THE 1960 ACT

For use by the applicant	For use by the International Bureau
This international application contains the following number of continuation sheets:	Registration No.: ....
Reference:	Filing date: .... Color: ☐

1 APPLICANT

(a) Name: ....

(b) Address: ....

....

(c) Telephone: .... Fax: ....

E-mail address: ....

☐ If there are several applicants, check box and use a continuation sheet, giving the above required information for each applicant

2 ADDRESS FOR CORRESPONDENCE
(where there are **several applicants** with different addresses and no representative is appointed, an address for correspondence should be indicated. Where no such address has been indicated, the address of the person named above shall be treated as the address for correspondence. If there is **only one applicant** and no representative is appointed, this item should be completed only if the address for correspondence is different from the address given in item 1(b))

Address for correspondence: ....

....

3 ENTITLEMENT TO FILE
(items (a), (b) and (c) must all be completed – even if the Contracting Party concerned is the same in all cases. Only one Contracting Party may be indicated with respect to each item. If any item is not applicable, write "None". A list of the Contracting Parties bound by the 1999 Act, the 1960 Act and/or the 1934 Act is attached to the present form)

(a) Contracting Party of which the applicant is a national: ....

(b) Contracting Party in which the applicant has a domicile: ....

(c) Contracting Party in which the applicant has a real and effective industrial or commercial establishment:

....

(d) Contracting Party in which the applicant has a habitual residence (only a Contracting Party bound by the **1999 Act** may be indicated): ....

☐ If there are several applicants, check box and use a continuation sheet, giving the above required information for each applicant

4 APPLICANT'S CONTRACTING PARTY
(indicate the Contracting Party, if any, bound by the *1999 Act* and mentioned in item 3, that is to be considered as the applicant's Contracting Party)

Applicant's Contracting Party: ....

☐ If there are several applicants, check box and use a continuation sheet, giving the above required information for each applicant

DM/1 (E) – IX.04

5 APPOINTMENT OF A REPRESENTATIVE (optional)

(a) Name: ....

(b) Address: ....

....

Telephone: .... Fax: ....

E-mail address: ....

(c) To appoint a representative, the present international application must either be signed by the applicant or be accompanied by a power of attorney (check the appropriate box):

☐ item 14 of the international application is signed by the applicant

☐ a power of attorney is attached to the present form

6 IDENTITY OF THE CREATOR
(this information is necessary if Bulgaria or Iceland is designated in item 10)

(a) Name: ....

(b) Address: ....

....

7 NUMBER OF INDUSTRIAL DESIGN(S), REPRODUCTION(S) AND/OR SPECIMEN(S)

(a) Total number of industrial designs: ....

(b) Total number of reproductions: ....

(i) in black and white: ....

(ii) in color: ....

(c) Total number of A4 pages comprising reproductions: ....

(d) Total number of specimens (if any): ....

8 PRODUCTS WHICH CONSTITUTE THE INDUSTRIAL DESIGN OR IN RELATION TO WHICH IT IS TO BE USED
(see Note 1 on cover page)

1. <u>Class</u> to which the industrial design(s) belong(s): ....
2. If the industrial design consists of a *product*, indicate its usual generic name, for example: 'chair'. If the industrial design consists of a *decorative motif*, indicate the product(s) in relation to which it is to be used, for example: 'drawing to be used with dishes' or 'motif for textiles'.

No. of the design (in numerical order)	Product(s)	Subclass (optional)

☐ If the space provided is not sufficient, check this box and use a continuation sheet.

DM/1 (E) – IX.04

9 DESCRIPTION (optional)
(only the characteristic features that appear in a reproduction should be described. The description should not disclose technical features of the operation of the industrial design or its possible utilization. If the description exceeds 100 words, an additional fee of 2 Swiss francs per word exceeding 100, shall be payable)

....

....

....

....

....

☐ If the space provided is not sufficient, check this box and use a continuation sheet.

10 DESIGNATED CONTRACTING PARTIES

Indicate, by ticking the appropriate box, each Contracting Party where protection is sought. The designated Contracting Party must be bound by an Act – the 1999 Act, the 1960 Act and/or 1934 Act – to which one of the Contracting Parties indicated in item 3 is also bound (a list of Contracting Parties bound by the 1999 Act, 1960 Act and/or 1934 Act is annexed to the present form):

☐ **AN** Netherlands Antilles
☐ **BG** Bulgaria
☐ **BJ** Benin
☐ **BX** Benelux
☐ **BZ** Belize
☐ **CH** Switzerland
☐ **CI** Côte d'Ivoire
☐ **DE** Germany
☐ **EE** Estonia
☐ **EG** Egypt
☐ **ES** Spain
☐ **FR** France
☐ **GA** Gabon
☐ **GE** Georgia
☐ **GR** Greece
☐ **HR** Croatia
☐ **HU** Hungary
☐ **ID** Indonesia
☐ **IS** Iceland[1]
☐ **IT** Italy
☐ **KG** Kyrgyzstan
☐ **KP** Democratic People's Republic of Korea
☐ **LI** Liechtenstein
☐ **MA** Morocco
☐ **MC** Monaco
☐ **MD** Rep. of Moldova
☐ **MK** The former Yugoslav Rep. of Macedonia
☐ **MN** Mongolia
☐ **NA** Namibia
☐ **NE** Niger
☐ **RO** Romania
☐ **SI** Slovenia
☐ **SN** Senegal
☐ **SR** Suriname
☐ **TN** Tunisia
☐ **UA** Ukraine
☐ **VA** Holy See
☐ **YU** Serbia and Montenegro

Others:

....

[1] If **Iceland** is designated, it is compulsory to indicate, in item 6, the identity of the creator. The latter declares that he believes himself to be the creator of the industrial design. Where the person identified as the creator is a person other than the applicant, it is hereby stated that the present international application has been assigned by the creator to the applicant.

11 PRIORITY CLAIM (optional)

☐ If the space provided is not sufficient, check this box and use a continuation sheet

☐ The applicant claims the priority of the earlier filing mentioned below

Indicate the number of each industrial design for which the priority is, or is not claimed. If no industrial design is indicated, it will be understood that the priority claim relates to all industrial designs included in the present application:

☐ Priority **is claimed** for the designs indicated below **or** ☐ Priority **is not claimed** for the designs indicated below

Office of earlier filing	No. of earlier filing (if available)	Date of earlier filing (dd/mm/yyyy)	No. of the industrial design(s)

DM/1 (E) – IX.04

12 INTERNATIONAL EXHIBITION (optional)

☐ If the space provided is not sufficient, check this box and use a continuation sheet

☐ The applicant claims that one or more designs were shown at an official, or officially recognized, international exhibition

(a) Place where exhibition was held: ....

(b) Name of the exhibition: ....

(c) Date on which product(s) was (were) first exhibited: ....

(d) Number of each industrial design shown, or not shown, at the exhibition concerned (if no industrial design is indicated, it will be understood that all industrial designs were shown at the above indicated exhibition):

☐ The industrial designs indicated below were **shown** at the exhibition concerned **or** ☐ The industrial designs indicated below were **not shown** at the exhibition concerned

....

13 PUBLICATION OF THE INTERNATIONAL REGISTRATION

(a) The applicant requests the publication in color ☐

(b) Timing of publication (publication will take place six months after the date of the international registration, unless the applicant requests one of the options below):

(i) The applicant requests the immediate publication of the international registration ☐

(ii) The applicant requests a deferment of publication ☐

- Period of deferment requested (in months): ....

Warning:

The period of deferment of publication cannot exceed 30 months counted from the date of the international application, or if priority is claimed, from the priority date. However:

¬if **Iceland**, **Indonesia, Netherlands Antilles**, **Tunisia** or **Holy See** is designated, or if **Ukraine** is designated under the 1999 Act, the applicant may NOT request deferment of publication;

¬if a Contracting Party is designated under the 1960 Act, or if **Croatia**, **Estonia** or **Slovenia** is designated, the period of deferment cannot exceed **12 months**.

14 SIGNATURE BY THE APPLICANT OR HIS REPRESENTATIVE

Applicant	Representative of the applicant
Name:	Name: ....
Signature and/or seal:	Signature and/or seal: ....
Date of signature (dd/mm/yyyy):	Date of signature (dd/mm/yyyy): ....

Name of the person to contact, if necessary: ....

15 OFFICE PRESENTING THE REQUEST (if applicable)

Name of the Office: ....

Date of receipt of the international application by the Office: ....

Signature and/or seal of the Office: ....

DM/1 (E) – IX.04

PAYMENT OF FEES

☐ The applicant has requested a deferment in publication (item 13(b)(ii)) and wishes to pay the publication fees at a later date but not later than three months before the period of deferment expires.

1. INSTRUCTION TO DEBIT FROM A CURRENT ACCOUNT
(if this box is completed, it is not necessary to complete items 2 and 3 below)

The International Bureau is hereby instructed to debit the required amount of fees from the following current account opened with the International Bureau:

Holder of the account: .... Account number: ....

Identity of the party giving the instruction: ....

2. METHOD OF PAYMENT

Identity of the party effecting the payment: ....

Payment received and acknowledged by WIPO	☐	WIPO receipt number	
Payment made by banker's cheque	☐	Cheque identification	dd/mm/yyyy ....
Payment made to WIPO bank account No. CH35 0425 1048 7080 8100 0 Credit Suisse, CH-1211 Geneva 70 Swift: CRESCH ZZ12A	☐	Payment identification	dd/mm/yyyy ....
Payment made to WIPO postal cheque account N° 12-5000-8, Geneva	☐	Payment identification	dd/mm/yyyy ....

GRAND TOTAL IN SWISS FRANCS (see note 3 or use the fee calculation sheet attached herewith) ....

FEE CALCULATION SHEET

This sheet is provided for the convenience of users.
It is not necessary to complete it if the fee calculator has been used.

3. AMOUNT OF FEES

(a) Basic fees

for one design 397.--

for each additional design
19 Swiss francs × .... designs in addition to the first ....

(b) Designation fees

(i) Standard designation fees (payable only for Contracting Parties designated under the 1999 or 1960 Act and for which an individual designation fee is not payable)

for first design
42 Swiss francs × .... Contracting Parties ....

for each additional design
2 Swiss francs × .... designs in addition to the first × .... Contracting Parties ....

(ii) Individual designation fee (payable only for the following Contracting Parties designated under the 1999 Act or the 1960 Act)

Bulgaria: 124 Swiss francs for first design + 32 Swiss francs × .... designs in addition to the first ....

Hungary: 126 Swiss francs for first design + 30 Swiss francs × .... designs in addition to the first ....

Kyrgyzstan: 188 Swiss francs for first design + 94 Swiss francs × .... designs in addition to the first ....

Rep. of Moldova: 65 Swiss francs for first design + 13 Swiss francs × .... designs in addition to the first ....

Romania: 101 Swiss francs for first design + 25 Swiss francs × .... designs in addition to the first ....

(c) Publication fees

(i) Fee per reproductions
12 Swiss francs × .... reproductions to be published in black and white ....

75 Swiss francs × .... reproductions to be published in color ....

(ii) Fee per page on which reproductions are presented
150 Swiss francs × .... page(s) in addition to the first ....

(d) Additional fee where the description exceeds 100 words

2 Swiss francs × .... words exceeding 100 ....

GRAND TOTAL (SWISS FRANCS) ....

DM/1 (E) – IX.04

CONTRACTING PARTIES OF THE HAGUE AGREEMENT

Contracting Parties of the Hague Agreement		Acts of the Hague Agreement		
		1999 Act	*1960 Act*	*1934 Act*
(AN)	Netherlands Antilles			√
(BG)	Bulgaria		√	
(BJ)	Benin		√	√
(BX)	Benelux		√	
(BZ)	Belize		√	
(CH)	Switzerland	√	√	√
(CI)	Cote D'Ivoire		√	√
(DE)	Germany		√	√
(EE)	Estonia	√		
(EG)	Egypt	√		√
(ES)	Spain	√		√
(FR)	France		√	√
(GA)	Gabon		√	
(GE)	Georgia	√	√	
(GR)	Greece		√	
(HR)	Croatia	√	√	
(HU)	Hungary	√	√	√
(ID)	Indonesia			√
(IS)	Iceland	√		
(IT)	Italy		√	
(KG)	Kyrgystan	√	√	
(KP)	Democratic People's Republic of Korea		√	
(LI)	Liechtenstein	√	√	√
(MA)	Morocco		√	√
(MC)	Monaco		√	√
(MD)	Republic of Moldova	√	√	
(MK)	The former Yugoslav Republic of Macedonia		√	
(MN)	Mongolia		√	
(NA)	Namibia	√		
(NE)	Niger		√	
(RO)	Romania	√	√	
(SI)	Slovenia	√	√	
(SN)	Senegal		√	√
(SR)	Suriname		√	√
(TN)	Tunisia			√
(UA)	Ukraine	√	√	
(VA)	Holy See			√
(YU)	Serbia & Montenegro		√	

APPENDIX IX

Contracting Parties of the Hague Agreement

7. Hague Agreement Concerning the International Registration of Industrial Designs[1]

Hague Agreement (1925), revised at London (1934) and The Hague (1960)[2] (supplemented by the Additional Act of Monaco (1961),[3] the Complementary Act of Stockholm (1967) and the Protocol of Geneva (1975),[4] and amended in 1979) and at Geneva (1999) (Hague Union)

Status on September 24, 2004

State	Date on which State became party to the Agreement	Date on which State became party to the London Act	Date on which State became party to the Hague Act[2]	Date on which State became party to the Complementary Act of Stockholm	Date on which State became party to the Geneva Act
Belgium[5,6]	April 1, 1979	–	August 1, 1984	May 28, 1979	–
Belize	July 12, 2003	–	July 12, 2003	July 12, 2003	–
Benin	November 2, 1986	November 2, 1986	November 2, 1986	January 2, 1987	–
Bulgaria	December 11, 1996	–	December 11, 1996	December 11, 1996	–
Côte d'Ivoire	May 30, 1993	May 30, 1993	May 30, 1993	May 30, 1993	–
Croatia	February 12, 2004	–	February 12, 2004	February 12, 2004	April 12, 2004
Democratic People's Republic of Korea	May 27, 1992	–	May 27, 1992	May 27, 1992	–
Egypt	July 1, 1952	July 1, 1952	–	–	August 27, 2004
Estonia	December 23, 2003	–	–	–	December 23, 2003
France[7]	October 20, 1930	June 25, 1939	August 1, 1984	September 27, 1975	–
Gabon	August 18, 2003	–	August 18, 2003	August 18, 2003	–
Georgia	August 1, 2003	–	August 1, 2003	August 1, 2003	December 23, 2003
Germany	June 1, 1928	June 13, 1939	August 1, 1984	September 27, 1975	–
Greece	April 18, 1997	–	April 18, 1997	April 18, 1997	–
Holy See	September 29, 1960	September 29, 1960	–	–	–
Hungary[8]	April 7, 1984	April 7, 1984[9]	August 1, 1984	April 7, 1984	May 1, 2004
Iceland	December 23, 2003	–	–	–	December 23, 2003
Indonesia	December 24, 1950	December 24, 1950	–	–	–
Italy	June 13, 1987	–	June 13, 1987	August 13, 1987	–
Kyrgyzstan	March 17, 2003	–	March 17, 2003	March 17, 2003	December 23, 2003
Liechtenstein	July 14, 1933	January 28, 1951	August 1, 1984	September 27, 1975	December 23, 2003[10]
Luxembourg[6]	April 1, 1979	–	August 1, 1984	May 28, 1979	–
Monaco	April 29, 1956	April 29, 1956	August 1, 1984	September 27, 1975	–
Mongolia	April 12, 1997	–	April 12, 1997	April 12, 1997	–
Morocco	October 20, 1930	January 21, 1941	October 13, 1999	October 13, 1999	–
Namibia	June 30, 2004	–	–	–	June 30, 2004
Netherlands[5,6]	April 1, 1979	–	August 1, 1984[11]	May 28, 1979[11]	–
Niger	September 20, 2004	–	September 20, 2004	September 20, 2004	–
Republic of Moldova	March 14, 1994	–	March 14, 1994	March 14, 1994	December 23, 2003[12]
Romania	July 18, 1992	–	July 18, 1992	July 18, 1992	December 23, 2003
Senegal	June 30, 1984	June 30, 1984	August 1, 1984	June 30, 1984	–
Serbia and Montenegro	December 30, 1993	–	December 30, 1993	December 30, 1993	–
Slovenia	January 13, 1995	–	January 13, 1995	January 13, 1995	December 23, 2003
Spain	June 1, 1928	March 2, 1956	–	–	December 23, 2003[13]
Suriname	November 25, 1975	November 25, 1975	August 1, 1984	February 23, 1977	–
Switzerland	June 1, 1928	November 24, 1939	August 1, 1984	September 27, 1975	December 23, 2003
The former Yugoslav Republic of Macedonia	March 18, 1997	–	March 18, 1997	March 18, 1997	–
Tunisia	October 20, 1930	October 4, 1942	–	–	–
Ukraine	August 28, 2002	–	August 28, 2002	August 28, 2002	December 23, 2003
(Total: 39 States)	(39)	(16)	(31)	(31)	(15)

- 15 -

7. Hague Agreement Concerning the International Registration of Industrial Designs[1]

Hague Agreement (1925), revised at London (1934) and The Hague (1960)[2] (supplemented by the
Additional Act of Monaco (1961),[3] the Complementary Act of Stockholm (1967)
and the Protocol of Geneva (1975),[4] and amended in 1979)
and at Geneva (1999)
(Hague Union)

(continued)

---

[1] The Geneva (1999) Act of the Hague Agreement Concerning the International Registration of Industrial Designs was adopted on July 2, 1999. The Geneva Act entered into force on December 23, 2003.

[2] The Protocol to the Hague Act (1960) is not yet in force. It has been ratified by or acceded to by the following States: Belgium, France, Germany, Italy, Liechtenstein, Monaco, Morocco, Netherlands, Switzerland.

[3] The Additional Act of Monaco (1961) is in force in respect of the following States as from the dates indicated: France (December 1, 1962), Germany (December 1, 1962), Liechtenstein (July 9, 1966), Monaco (September 14, 1963), Netherlands (as far as the Netherlands Antilles is concerned) (September 14, 1963), Spain (August 31, 1969), Suriname (November 25, 1975) and Switzerland (December 21, 1962). See also footnote 5.

[4] The Protocol of Geneva (1975), in accordance with Article 11(2)(a) thereof, ceased to have effect as of August 1, 1984; however, as provided by Article 11(2)(b), States bound by the Protocol (Belgium (as from April 1, 1979), France (as from February 18, 1980), Germany (as from December 26, 1981), Hungary (as from April 7, 1984), Liechtenstein (as from April 1, 1979), Luxembourg (as from April 1, 1979), Monaco (as from March 5, 1981), Netherlands (as from April 1, 1979), Senegal (as from June 30, 1984), Suriname (as from April 1, 1979) and Switzerland (as from April 1, 1979)) are not relieved of their obligations thereunder in respect of industrial designs whose date of international deposit is prior to August 1, 1984.

[5] Belgium had withdrawn from the Hague Union with effect from January 1, 1975. The Netherlands had denounced, in respect of the Kingdom in Europe and with effect from January 1, 1975, the Hague Agreement (1925) and the subsequent Acts to which the Netherlands had adhered, specifying that the said Agreement and Acts – London Act (1934) and Additional Act of Monaco (1961) – would remain in force in respect of the Netherlands Antilles and Suriname. As a result of their ratification of the Protocol of Geneva (1975) and its entry into force on April 1, 1979, Belgium and the Netherlands became, again, as from that date, members of the Hague Union.

[6] The territories in Europe of Belgium, Luxembourg and the Netherlands are, for the application of the Hague Agreement, to be deemed a single country.

[7] Including all Overseas Departments and Territories.

[8] With the declaration that Hungary does not consider itself bound by the Protocol annexed to the Hague Act (1960).

[9] The London Act will cease to be effective in respect of Hungary as of February 1, 2005.

[10] With the declaration, in accordance with Article 30(1) of the Geneva (1999) Act, that, under Article 17(3)(c) of the said Act, the maximum duration of protection provided for by the law of the Principality of Liechtenstein is 25 years.

[11] Ratification for the Kingdom in Europe.

[12] With the declaration, in accordance with Article 30(1) of the Geneva (1999) Act, that, under Article 7(2) of the said Act, the prescribed designation fee referred to in Article 7(1) of the said Act shall be replaced by an individual designation fee.

[13] With the declaration, in accordance with Rule 18(1)(b) of the Common Regulations under the Geneva (1999) Act, the 1960 Act and the 1934 Act of the Hague Agreement, that the prescribed period of six months for the notification of refusal of the effects of an international registration shall be replaced by a period of 12 months; and with the declaration, under Rule 18(1)(c)(i) of the said Regulations, that the international registration shall produce the effect referred to in Article 14(2)(a) of the Geneva (1999) Act at the latest six months after the date referred to in that Article.

APPENDIX X

International Registration Bulletin Publication of Multiple Designs Registration

(11) **DM/032 604** (15) **10.11.1994**

(73) PHILIPS ELECTRONICS N.V., 1, Groenewoudseweg, NL-5621 BA Eindhoven (NL) – (86) (87) (88) NL – (74) Internationaal Octrooibureau B.V., 6, Prof. Holstlaan, NL-5656 AA Eindhoven (NL) – (28) 2 – (54) Dry shavers / *Rasoirs électriques* – (51) Cl. 28-03.

(81) I. BX, CH, DE, FR, IT.
(45) 10.05.1995.

1.1

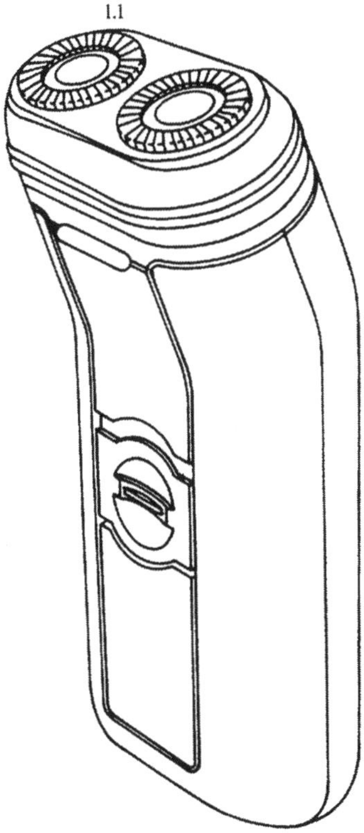

1.2

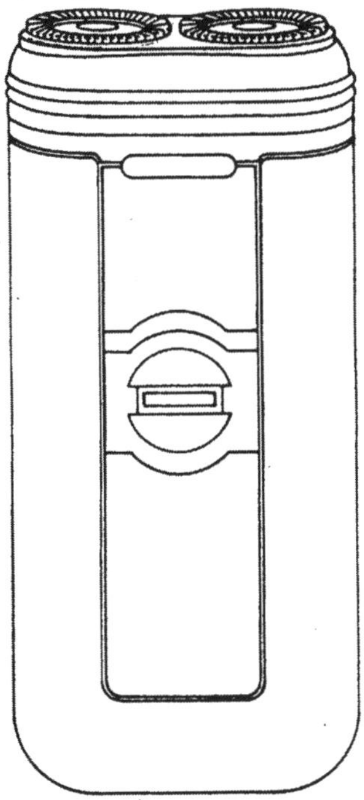

DM/032 604 (suite/cont'd.)

1.3

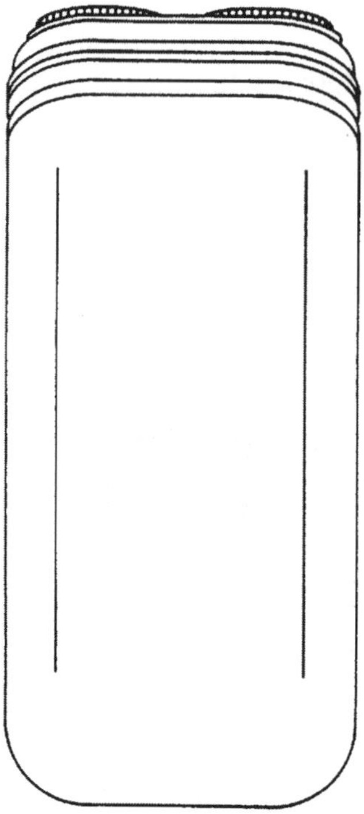

1.4

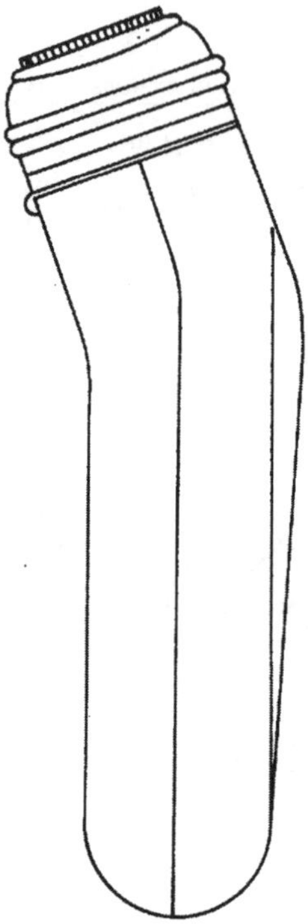

1.5

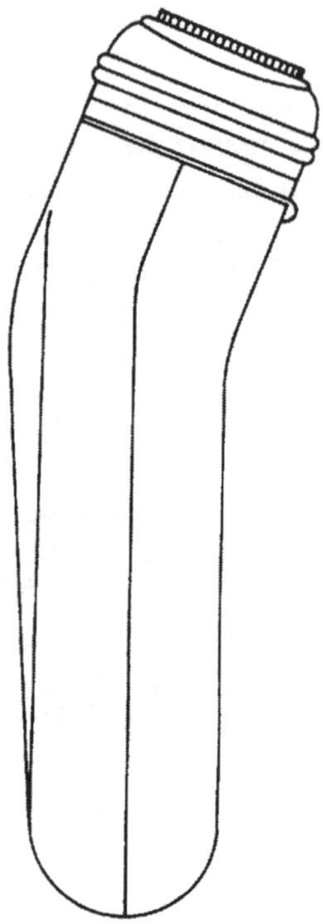

1.6

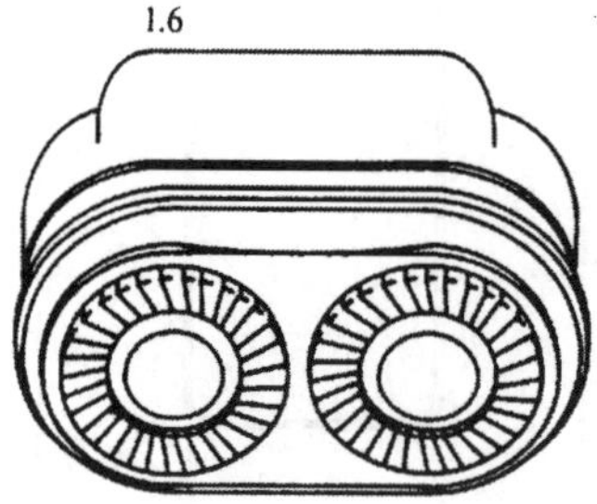

1.7

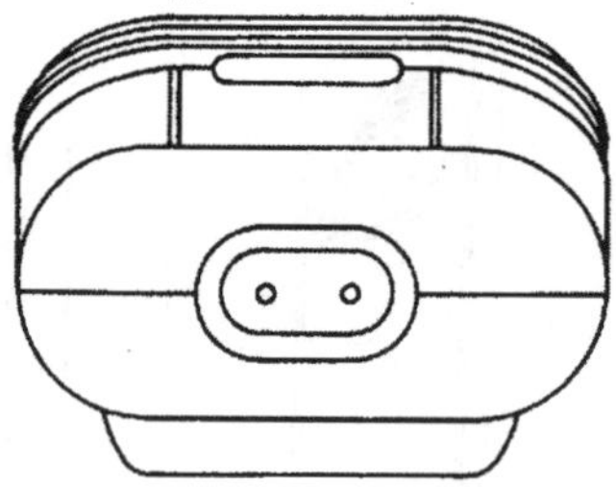

DM/032 604 (suite/cont'd.)

2.1

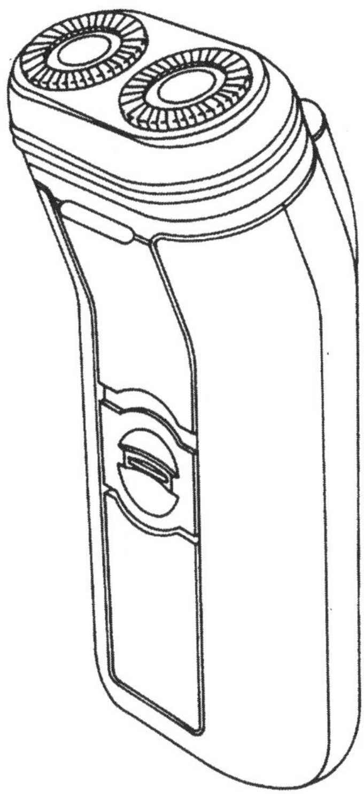

2.2

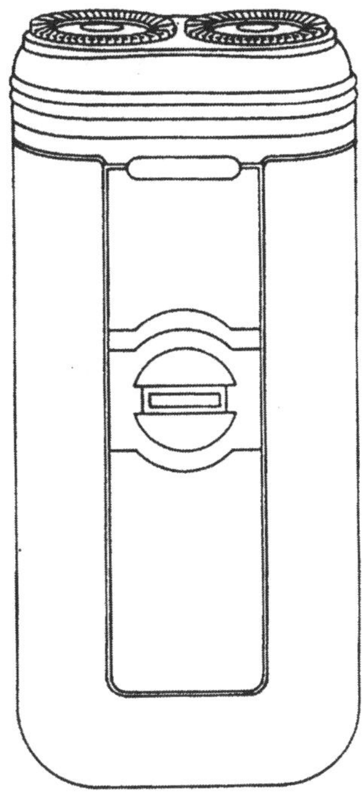

DM/032 604 (suite/cont'd.)

2.3

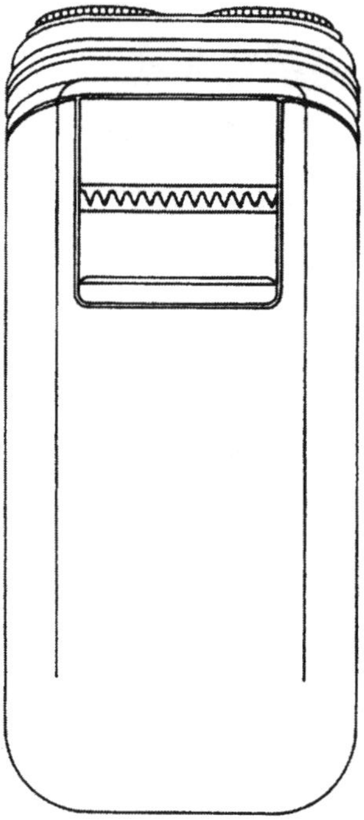

2.4

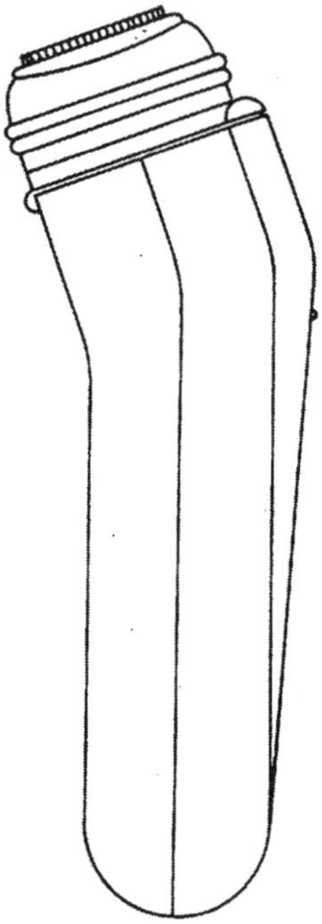

2.5

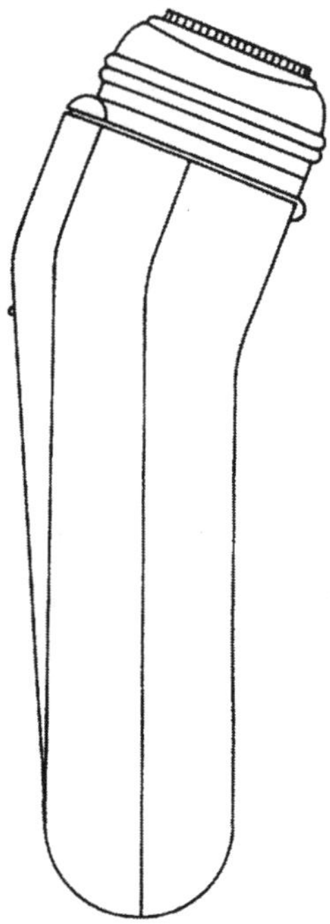

2.6

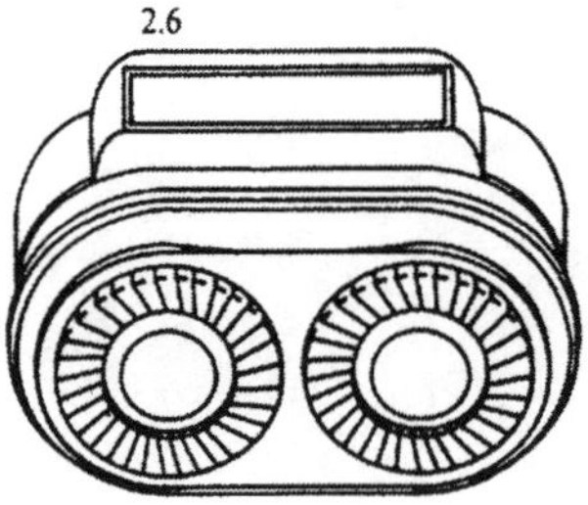

2.7

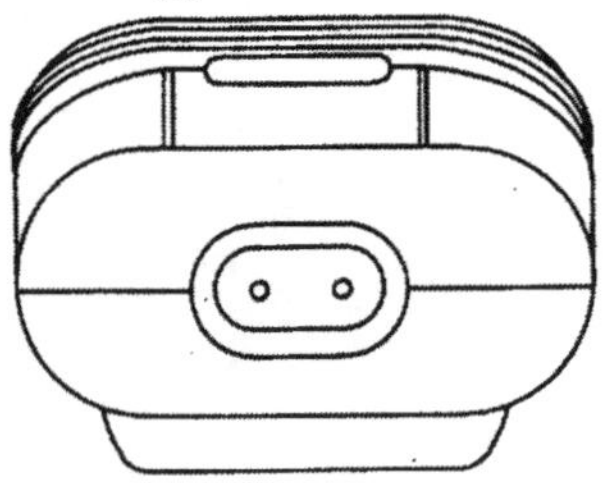

APPENDIX XI

International Classification (Locarno Agreement) List of Classes

CLASS 1 FOODSTUFFS

CLASS 2 ARTICLES OF CLOTHING AND HABERDASHERY

CLASS 3 TRAVEL GOODS, CASES. PARASOLS AND PERSONAL BELONGINGS, NOT ELSEWHERE SPECIFIED

CLASS 4 BRUSHWARE

CLASS 5 TEXTILE PIECEGOODS, ARTIFICIAL AND NATURAL SHEET MATERIAL

CLASS 6 FURNISHING

CLASS 7 HOUSEHOLD GOODS, NOT ELSEWHERE SPECIFIED

CLASS 8 TOOLS AND HARDWARE

CLASS 9 PACKAGES AND CONTAINERS FOR THE TRANSPORT OR HANDLING OF GOODS

CLASS 10 CLOCKS AND WATCHES AND OTHER MEASURING INSTRUMENTS, CHECKING AND SIGNALLING INSTRUMENTS

CLASS 11 ARTICLES OF ADORNMENT

CLASS 12 MEANS OF TRANSPORT OR HOISTING

CLASS 13 EQUIPMENT FOR PRODUCTION, DISTRIBUTION OR TRANSFORMATION OF ELECTRICITY

CLASS 14 RECORDING, COMMUNICATION OR INFORMATION RETRIEVAL EQUIPMENT

CLASS 15 MACHINES, NOT ELSEWHERE SPECIFIED

CLASS 16 PHOTOGRAPHIC, CINEMATOGRAPHIC AND OPTICAL APPARATUS

CLASS 17 MUSICAL INSTRUMENTS

CLASS 18 PRINTING AND OFFICE MACHINERY

CLASS 19 STATIONERY AND OFFICE EQUIPMENT, ARTISTS' AND TEACHING MATERIALS

CLASS 20 SALES AND ADVERTISING EQUIPMENT, SIGNS

CLASS 21 GAMES, TOYS, TENTS AND SPORTS GOODS

CLASS 22 ARMS, PYROTECHNIC ARTICLES, ARTICLES FOR HUNTING, FISHING AND PEST KILLING

CLASS 23 FLUID DISTRIBUTION EQUIPMENT, SANITARY, HEATING, VENTILATION AND AIR-CONDITIONING EQUIPMENT, SOLID FUEL

CLASS 24 MEDICAL AND LABORATORY EQUIPMENT

CLASS 25 BUILDING UNITS AND CONSTRUCTION ELEMENTS

CLASS 26 LIGHTING APPARATUS

CLASS 27 TOBACCO AND SMOKERS' SUPPLIES

CLASS 28 PHARMACEUTICAL AND COSMETIC PRODUCTS, TOILET ARTICLES AND APPARATUS

CLASS 29 DEVICES AND EQUIPMENT AGAINST FIRE HAZARDS, FOR ACCIDENT PREVENTION AND FOR RESCUE

CLASS 30 ARTICLES FOR THE CARE AND HANDLING OF ANIMALS

CLASS 31 MACHINES AND APPLIANCES FOR PREPARING FOOD OR DRINK. NOT ELSEWHERE SPECIFIED

CLASS 99 MISCELLANEOUS

Index

Abbreviations:

DC – Diplomatic Conference

GA – Geneva Act (1999)

IR – Geneva Act International Registration

ME – Meeting of Experts